DONALD VOET

University of Pennsylvania

JUDITH G. VOET

Swarthmore College

BIOCHEMISTRY

SECOND EDITION

1996 Supplement

JOHN WILEY & SONS, INC.
New York   Chichester   Brisbane
Toronto   Singapore

ISBN 0-471-12414-1

Printed in the United States of America

10 9 8 7 6 5 4 3 2 1

Printed and bound by Port City Press

INTRODUCTION

This is the first annual Supplement to *Biochemistry* (2nd ed.) by Donald Voet and Judith G. Voet. The biochemical literature covered in this Supplement extends from the time that of *Biochemistry* (2nd ed.) went to press, around September, 1994, to around April, 1996. During this ~18 month time period, the biochemical literature has continued its explosive growth. It is therefore increasingly important for the student and teacher alike to keep up with the literature. The annual Supplements to *Biochemistry* should be taken as a guide for doing so.

As in the three Supplements to the First Edition of *Biochemistry,* this Supplement is keyed to the Second Edition in that we refer to new advances in terms of the textbook sections in which they would logically fit. Since space limitations permit only the most cursory discussions of these topics, the interested reader should consult the pertinent references provided at the end of each discussion. References that are not preceded by a discussion or which are placed under the subheading ''Additional References'' provide up-to-date coverage of subjects discussed in the corresponding section of the textbook but which are not otherwise treated in this Supplement.

The reader should note that two types of ancillary materials to accompany *Biochemistry* (2nd ed.) have been published recently:

- A set of two CD-ROMs containing nearly all of the figures in the Second Edition, one in a form suitable for projection and the other in a form suitable for printing (ISBN 0-471-05882-3).

- A diskette containing a series of kinemages in 22 Exercises for the interactive display of various proteins and nucleic acids (ISBN 0-471-13558-5 for IBM-compatible computers and ISBN 0-471-13559-3 for Macintosh computers). We plan to develop several additional Exercises which will be made available over the World Wide Web at http://www.wiley.com/Guides/Chemistry/Voet.html

In addition, a *Student's Companion to Biochemistry* by Akif Uzman, Joseph Eichberg, William Widger, Donald Voet, and Judith G. Voet will be published in the Fall of 1996. It contains additional discussions, numerous new problems, and their detailed answers.

If you purchased this Supplement or *Biochemistry* (2nd ed.) directly from John Wiley & Sons, Inc., your subscription for this update service has been recorded. If, however, you purchased this Supplement from a bookstore and wish to receive future Supplements billed separately, with a 30-day examination review, please send your name, company name (if applicable), address, and the notation ''SUBSCRIPTION SERVICE FOR VOET/VOET: BIOCHEMISTRY, Second Edition'' to: Supplement Department, John Wiley & Sons, Inc., One Wiley Drive, Somerset, NJ 08875, Tel. 1-800-225-5945, FAX: 908-302-2300. For customers outside the United States, please contact the Wiley office nearest you at the addresses given on the bottom of page vi of the textbook.

Donald Voet
Judith G. Voet

Chapter 6

COVALENT STRUCTURES OF PROTEINS

1. PRIMARY STRUCTURE DETERMINATION

Biemann, K. and Papayannopouos, I.A., Amino acid sequencing of proteins, *Acc. Chem. Res.* **27,** 370–378 (1994). [Discusses mass spectrometric techniques.]

Gross, M.L., Tandem mass spectrometric strategies for determining structure of biologically interesting molecules, *Acc. Chem. Res.* **27,** 361–369 (1994).

Handcock, W.S. (Ed.), *New Methods in Peptide Mapping for the Characterization of Proteins,* CRC Press (1996).

Mann, M. and Matthias, W., Electrospray mass spectrometry for protein characterization, *Trends Biochem. Sci.* **20,** 219–224 (1995).

Stults, J.T., Matrix-assisted laser desorption/ionization mass spectrometry (MALDI-MS), *Curr. Opin. Struct. Biol.* **5,** 691–698 (1995).

2. PROTEIN MODIFICATION

Lundblad, R.L., *Techniques in Protein Modification,* CRC Press (1995). [A laboratory support book that describes the most frequently used methods for the site-specific chemical modification of proteins.]

3. CHEMICAL EVOLUTION

(a) The Divergence Times of the Major kingdoms of Living Organisms as Determined from Amino Acid Sequence Comparisons

Estimates of when two species diverged, that is, when they last had a common ancestor, are based largely on the radiodated fossil record. However, the macrofossil record only extends back ~600 million years (after multicellular organisms arose) and phylogenetic comparisons of microfossils (fossils of single-celled organisms) are unreliable. Thus, previous estimates of when the major groupings of organisms (animals, plants, fungi, protozoa, eubacteria, and archaebacteria; Fig. 1-11) diverged from one another (e.g., the right side of Fig. 6-15) are only approximations based mainly on the consideration of shared characteristics.

The burgeoning data base of amino acid sequences has permitted comparisons of the sequences of a large variety of enzymes that each have homologous representatives in many of the above major groupings (531 sequences comprising 57 different enzymes). Analysis of these data indicates the existence of a ''protein clock'' that can provide reliable estimates of when these groupings diverged. This "protein clock," which is based on the supposition that homologous sequences diverge at a uniform rate, was calibrated using sequences from vertebrates for which there is a reasonably reliable fossil record. The results of this analysis reveal that animals, plants, and fungi last had a common ancestor ~1 billion years ago with plants having diverged from animals slightly before fungi; that the major protozoan lineages separated from those of other eukaryotes ~1.2 billion years ago; that eukaryotes last shared a common ancestor with archaebacteria ~1.8 billion years ago and with eubacteria slightly more than 2 billion years ago; and that gram-positive and gram-negative eubacteria diverged ~1.4 billion years ago.

Doolittle, R.F., Feng, D.-F., Tsang, S., Cho, G., and Little, E., Determining divergence times of the major kingdoms of living organisms with a protein clock, *Science* **271,** 470–477 (1996).

(b) Additional References

Doolittle, R.F., The multiplicity of protein domains, *Annu. Rev. Biochem.* **64,** 287–314 (1995).

Chapter 7

THREE DIMENSIONAL STRUCTURES OF PROTEINS

2. FIBROUS PROTEINS

(a) Structure of Collagen

The X-ray structure of a collagen-like polypeptide has been determined at 1.9-Å resolution. The polypeptide has the sequence (Pro-Hyp-Gly)$_4$-(Pro-Hyp-Ala)-(Pro-Hyp-Gly)$_5$, where Hyp is the abbreviation for the 4-hydroxyproline residue [recall that collagen has the monotonously repeating sequence (Gly-X-Y)$_n$ where X is often Pro and Y is often Hyp]. Three such polypeptides associate to form a triple helical structure whose basic features closely resemble those of the collagen structure deduced from X-ray fiber diffraction studies (Figs. 7-31 and 7-32).

The X-ray structure additionally reveals that water plays a crucial role in stabilizing the triple helix. The 87-Å long cylindrical molecule is surrounded by a sheath of ordered water molecules that associate with the polypeptides and with each other. Water molecules are hydrogen bonded to the Gly and Hyp carbonyl oxygen atoms as well as to the 4-OH groups of the Hyp residues. Evidently, the 4-OH groups function to anchor this solvent network to the polypeptides. The water molecules hydrogen bond with each other in repetitive patterns so as to often form pentagonal rings similar to those in clathrates (Fig. 7-52).

The central Gly → Ala substitution in each polypeptide chain, which reduces the melting temperature (T_m) of the triple helix from 62°C to 29°C, locally distorts the collagen triple helix.

The need to accommodate the three additional methyl groups in the tightly packed interior of the triple helix pries apart the polypeptide chains in the region of the substitutions so as to rupture the hydrogen bonds that would otherwise link the main chain N–H group of each Ala (normally Gly) to the carbonyl oxygen of the adjacent Pro in a neighboring chain (Fig. 7-32). Rather, these hydrogen bonding groups are bridged by water molecules that insinuate themselves into the distorted part of the structure. Similar distortions are likely to occur in the Gly → X mutated collagens responsible for such diseases as osteogenesis imperfecta and Ehler-Danlos syndrome IV.

The structure of this collagen-like triple helical structure can be interactively examined via computer graphics in EXERCISE 3 of the KINEMAGES to accompany *BIOCHEMISTRY* 2/E by Donald Voet and Judith G. Voet.

Bella, J., Eaton, M., Brodsky, B. and Berman, H.M., Crystal and molecular structure of a collagen-like peptide at 1.9 Å resolution, *Science* **266,** 75–81 (1994).

Bella, J., Brodsky, B. and Berman, H.M., Hydration structure of a collagen peptide, *Structure* **3,** 893–906 (1995).

(b) Additional References

Fuchs, E., Keratins and the skin, *Annu. Rev. Cell Biol.* **11,** 123–153 (1995).

Parry, D. A. D., Hard α-keratin IF: A structural model lacking a head-to-tail molecular overlap but having hybrid features characteristic of both epidermal keratin and vimentin IF, *Proteins* **22,** 267–272 (1995).

Prokop, D.J. and Kivirikko, K.I., Collagens: Biology, disease, and potentials for therapy, *Annu. Rev. Biochem.* **64,** 403–434 (1995).

3. GLOBULAR PROTEINS

Darby, N.J. and Creighton, T.E., *Protein Structure,* IRL Press (1993). [A general overview.]

Kyte, J., *Structure in Protein Chemistry,* Garland (1995).

Shirley, B.A. (Ed.), *Protein Stability and Folding,* Humana Press (1995).

Wüthrich, K., NMR – This other method for protein and nucleic structure determination, *Acta Cryst.* **D51,** 249–270 (1995).

4. PROTEIN STABILITY

(a) Role of Salt Bridges

The strong attractive interaction between contacting (hydrogen bonded) oppositely charged side chains might naively suggest that such salt bridges (ion pairs) play an important structural role in proteins. This, however, is not the case because the energy gain of a salt bridge's charge-charge interaction relative to the unfolded state usually fails to compensate for the entropic penalty of localizing the salt bridge's charged side chains and the charged groups' increased solvation energy in the unfolded state. Indeed, a survey of 8 protein families comprising 57 proteins that have a total of 339 salt bridges revealed that surface-exposed salt bridges are poorly conserved. However, the degree of salt bridge conservation increases with the extent that they are buried beneath the surface of the protein. Thus, although buried salt bridges are less common than those that are surface-exposed, they appear to have a more important structural role.

An additional stabilizing aspect of buried salt bridges is that the dielectric constant in the interior of a protein is less than that on its surface and hence the charge-charge interactions of a buried salt bridge should be greater than those of a surface-exposed salt bridge. Nevertheless, calculations indicate that the energy of a buried charge-charge interaction is insufficient to compensate for the increased energy of solvation of these charged side chains in the unfolded protein. These calculations therefore predict that the stability of a native protein containing a buried salt bridge would be increased if the side chains forming the salt bridge were replaced by hydrophobic side chains of the same size and shape.

This hypothesis is supported by studies on the *arc* repressor of bacteriophage P22 (Section 29-3D). Wild-type *arc* repressor contains a salt bridge triad consisting of Arg 31, Glu 36, and Arg 40 in which Glu 36 is completely inaccessible to solvent. Mutagenesis generating all 8000 combinations of 20 amino acids at each of these 3 positions, followed by selection for P22 bacteriophage containing a functional *arc* repressor, yielded only four mutant repressors that are fully active: MYL, IYV, VYI, and VYV [where the one letter amino acid code (Table 4-1) is used here for residues at positions 31, 36, and 40, respectively]. Another 16 mutants proteins were found to be partially active; they contained mostly nonpolar residues at the mutated positions. Six of these fully or partially active variants, MYL, VYI, IYV, MWL, LMI, and QYV were purified and found to be more stable than the wild-type (RER) protein by 6.7 to 18.9 kJ/mol of the dimeric protein. Moreover, the X-ray structure of the MYL variant is essentially identical to that of the wild-type protein with the exception of the side chains of Met 31, Tyr 36 and Leu 40. These, however, pack against each other so as to efficiently fill the space occupied by the Arg 31–Glu 36–Arg 40 salt bridge in the wild-type protein. In contrast, replacing all three salt bridge residues with Ala (AAA) yields an inactive protein that is 15.9 kJ/mol less stable than wild-type protein. Thus, simple hydrophobic interactions can confer more stabilizing interactions than a buried salt bridge and yet provide similar conformational specificity.

Hendsch, Z. and Tidor, B., Do salt bridges stabilize proteins? A continuum electrostatic analysis, *Prot. Sci.* **3,** 211–226 (1994).

Scheuler, O. and Margelit, H., Conservation of salt bridges in protein families, *J. Mol. Biol.* **248,** 125–135 (1995)

Waldburger, C.D., Schildbach, J.F., and Sauer, R.T., Are buried salt bridges important for protein stability and conformational specificity, *Nature Struct. Biol.* **2,** 122–128 (1995).

(b) How Do Thermostable Proteins Differ from their Normal Homologs?

Certain species of bacteria known as **hyperthermophiles** grow at temperatures near 100°C (they live in such places as hot springs and submarine hydrothermal vents). These organisms have many of the same metabolic pathways as do **mesophiles** (organisms that grow at ''normal'' temperatures). Yet, most mesophilic proteins denature at the temperatures that hyperthermophiles thrive. What is the structural basis for the thermostability of hyperthermophilic proteins?

Comparisons of the X-ray structures of several hyperthermophilic enzymes with their heat-labile homologs have failed to reveal any striking differences between them. These proteins exhibit some variations in secondary structure but no more so than is often the case for homologous proteins from distantly related mesophiles. However, the recently determined structures of two thermostable proteins [glutamate dehydrogenase (GluDH; Section 24-1) from *Pyroccocus furiosus* and indol-3-glycerol phosphate synthase (IGPS; Section 24-5B) from *Sulfolobus solfaticarus]*, when compared with those of their mesophilic homologs (those from *Clostridium symbiosum* and from *E. coli* , respectively), have revealed one mechanism of thermostability. Both thermostable enzymes have a superabundance of salt bridges, many of which are arranged in extensive networks. Indeed, one of these networks in the *Pyrococcus* GluDH consists of 18 side chains.

The idea that salt bridges can stabilize a protein structure is apparently at odds with the conclusions of the preceding paragraph that the charge-charge interaction between the members of an ion pair yields less free energy than is lost on desolvating and immobilizing its component side chains. However, the key to this apparent paradox is that the salt bridges in thermostable proteins form networks. Thus, the gain in charge–charge free energy upon associating a third charged group with an ion pair is comparable to that between the members of this ion pair, whereas the free energy lost on desolvating and immobilizing the third side chain is only about half that lost in bringing together the first two side chains. The same is, of course, true for the addition of a fourth, fifth, etc. side chain to a salt bridge network.

Not all thermostable proteins contain such a high incidence of salt bridges. Structural comparisons suggest that these proteins are stabilized by a combination of small effects the most important of which are an increased size in the protein's hydrophobic core, an increased size in the interface between its domains and/or subunits, and a more tightly packed core as evidenced by a reduced surface-to-volume ratio.

The fact that the proteins of hyperthermophiles and mesophiles are homologous and carry out much the same functions indicates that mesophilic proteins are by no means maximally stable. This, in turn, strongly suggests that the marginal stability of most proteins under physiological conditions (averaging ~0.4 kJ/mol of amino acid residues) is an essential property that has arisen through evolutionary design. Perhaps this marginal stability eliminates otherwise stable non-native conformations and/or facilitates molecular chaperones in correcting such misfolded conformations (Section 8-1C) and in unfolding proteins to permit their insertion into or transport through membranes (Section 11-4C). Alternatively, the marginal stability of proteins may facilitate their programmed degradation (Section 30-6).

Goldman, A., How to make my blood boil, *Structure* **3**, 1277–1279 (1995).

Hennig, M., Darimont, B., Sterner, R., Kirschner, K., and Jansonius, J.N., 2.0 Å structure of indole-3-glycerol phosphate synthase from the hyperthermophile *Sulfolobus solfataricus:* Possible determinants of protein stability, *Structure* **3**, 1295–1306 (1995).

Rees, D.C. and Adams, M.W.W., Hyperthermophiles: Taking the heat and loving it, *Structure* **3**, 251–254 (1995).

Yip, K.S.P., *et al.,* The structure of *Pyrococcus furiosis* glutamate dehydrogenase reveals a key role for ion-pair networks in maintaining enzyme stability at extreme temperatures, *Structure* **3**, 1147–1158 (1995).

(c) Additional References

Bordo, D. and Argos, P., The role of side-chain hydrogen bonds in the formation and stabilization of secondary structure in soluble proteins, *J. Mol. Biol.* **243**, 504–519 (1994).

Honig, B. and Nichols, A., Classical electrostatics in biology and chemistry, *Science* **268**, 1144–1149 (1995).

Lazarides, T., Archontis, G., and Karplus, M., Enthalpic contribution to protein stability: Insights from atom-based calculations and statistical mechanics, *Adv. Prot. Chem.* **47**, 231–306 (1995).

Matthews, B.W., Studies on protein stability with T4 lysozyme, *Adv. Prot. Chem.* **46**, 249–278 (1995). [A distillation of the results of stability studies on a large number of mutant varieties of lysozyme from bacteriophage T4, many of which have also been structurally determined by X-ray analysis.]

Chapter 8

PROTEIN FOLDING, DYNAMICS, AND STRUCTURAL EVOLUTION

1. PROTEIN FOLDING: THEORY AND EXPERIMENT

(a) Structure of GroEL

The Hsp60 and Hsp10 chaperonins, which in *E. coli* are named GroEL and GroES, function together in an ATP-dependent process to facilitate the folding of proteins to their native conformations (Fig. 8-11). The X-ray structure of GroEL shows, in agreement with electron micrograph-derived images (e.g., Fig. 8-10), that GroEL's 14 identical subunits associate to form a porous thick-walled hollow cylinder that is slightly taller (146 Å) than it is wide (137 Å). It consists of two nearly 7-fold symmetric rings of subunits stacked back-to-back with 2-fold symmetry (approximate D_7 symmetry; Section 7-5B).

Each 547-residue subunit consists of three domains: a large equatorial domain (residues 1-135 and 410–547) that forms the waist of the protein and holds together its rings; a loosely structured apical domain (residues 191–376) that forms the open ends of the cylinder; and a small intermediate domain (residues 136–190 and 377-409) that connects the equatorial and apical domains. Neighboring subunits are partially out of contact, thereby forming elliptical side windows (36 × 13-Å) at the level of the intermediate domains that open into the central channel.

These side windows may function to provide ATP with access to its binding site (see below).

The central channel appears to have a uniform diameter of ~45 Å over most of its length but expands outwards to ~90 Å at the level of the intermediate domains. However, electron microscopy-based images show density obstructing the channel at its equatorial level. This equatorial obstruction is thought to be caused by each subunit's N-terminal 5 residues and the C-terminal 22 residues, which are not resolved in the X-ray structure. The obstruction, whose existence and molecular mass was confirmed by small-angle neutron scattering studies, is probably responsible for the observed inability of a folding polypeptide to transfer between the two rings of GroEL (see below).

The central cavity in each ring (toroid) has a volume sufficient to contain a folding protein of at least 35 kD, assuming that a compact folding intermediate occupies ~1.7 times the volume of the corresponding native protein. Nevertheless, the model most consistent with small-angle neutron scattering from the complex of GroEL with the monomeric 33.8-kD enzyme rhodanese (Fig. 8-19), places a single molecule of rhodanese at one opening of GroEL's central cavity, much like a cork in a champagne bottle. Electron microscope-based images of denatured malate dehydrogenase with GroEL show this protein to be similarly positioned.

Mutations that impair polypeptide binding to GroEL all map to a poorly resolved (and presumably flexible) segment of the apical domain that faces the central channel. In fact, changing any one of eight highly conserved hydrophobic residues in this region to Glu or Ser abolishes polypeptide binding. Thus, it seems likely that these residues provide the binding site(s) for non-native polypeptides. Interestingly, mutations of these same residues also abolish the binding of GroES.

The binding and hydrolysis of ATP causes GroEL to release its bound polypeptide (Fig. 8-11b). The X-ray structure of GroEL with ATPγS bound to each subunit (ATPγS is a poorly hydrolyzable analog of ATP in which S replaces one of the O atoms substituent to P_γ) indicates that ATP binds to a novel pocket in the equatorial domain that faces the central channel. The residues forming this pocket are highly conserved among chaperonins. The only significant differences between the structures of the GroEL–ATPγS complex and that of GroEL alone are modest movements of the residues in the vicinity of the ATP pocket. Thus, the nature of the allosteric changes to the apical domain when ATP binds to the equatorial domain of any subunit remain unknown.

Braig, K., Otwinowski, Z., Hegde, R., Bolsvert, D.C., Joachimiak, A., Horwich, A.L., and Sigler, P.B., The crystal structure of the bacterial chaperonin GroEL at 2.8 Å, *Nature* **371,** 578-586 (1994).

Braig, K. Adams, P.D., and Brünger, A.T., Conformatioal variability of the refined structure of the chaperonin GroEL at 2.8 Å resolution, *Nature Struct. Biol.* **2,** 1083–1094 (1995).

Thiyagarajan, P., Henderson, S.J., and Joachimiak, A., Solution structure of GroEL and its complex with rodanese from small-angle neutron scattering, *Structure* **4,** 79–88 (1996).

Boisvert, D.C., Wang, J., Otwinowski, Z., Horwich, A.L., and Sigler, P.B., The 2.4 Å crystal structure of the bacterial chaperonin GroEL complexed with ATPγS, *Nature Struct. Biol.* **3,** 170–177 (1996).

(b) X-Ray Structures of GroES and it Homolog Cpn10

The X-ray structure of GroES shows that this protein's 7 identical 97-residue subunits form an approximately 7-fold symmetric dome-like structure that is 30 Å high, 70 to 80 Å in diameter, and has a ~12-Å in diameter central opening in its roof (an architecture resembling that of the Pantheon in Rome). Each subunit consists of an irregular antiparallel β-barrel from which two β-hairpins project. One of these β-hairpins (residues 47-55) extends from the top of the β-barrel towards the protein's 7-fold axis where it interacts with the other such β-hairpins to form the roof of the dome. The other β-hairpin (residues 16-33) extends from the opposite side of the β-barrel outwards from the bottom outer rim of the dome. This so-called mobile loop is observed in only one of GroES's 7 subunits; it is apparently disordered in the other subunits in agreement with the results of NMR studies of uncomplexed GroES in solution. The X-ray structure of the 99-residue GroES homolog Cpn10 from *Mycobacterium leprae* (the organism causing leprosy) closely resembles that of GroES; its mobile loop is unobserved in all 7 of its subunits. The inner surface of the dome in both GroES and Cpn10 is lined with hydrophilic residues.

Hunt, J.F., Weaver, A.J., Landry, S.J., Gierasch, L., and Deisenhofer, J., The crystal structure of the GroES co-chaperonin at 2.8 Å resolution, *Nature* **379,** 37–45 (1996).

Mande, S.C., Mehra, V., Bloom, B.R., and Hol, W.G.J., Structure of the heat shock protein chaperonin-10 of *Mycobacterium leprae, Science* **271,** 203–207 (1996).

Saibil, H. The lid that shapes the pot: Structure and function of the chaperonin GroES, *Structure* **4,** 1–4 (1996).

(c) Mechanism of GroEL–GroES-Facilitated Protein Folding

Electron microscopy studies reveal that, in the GroEL–GroES complex, the GroES dome is positioned like a cap over the GroEL cylinder (as is drawn in Fig. 8-10*a*) and hence GroES's mobile loops must mediate this association. These studies further show that the apical domains of the GroEL toroid that is bound to the GroES open upward and outward like the petals of an opening flower to form, together with the GroES, a fully enclosed 70 Å in diameter dome-shaped cavity in which a protein can fold in an environment that prevents it from nonspecifically aggregating with other unfolded proteins.

Ongoing investigations have necessitated alterations in the mechanism of GroEL–GroES-facilitated protein folding that is diagrammed in Fig. 8-11*a*. The GroEL–GroES cycle is initiated by the binding of 7 ATPs to one toroid of GroEL, which subsequently binds GroES. This, in turn, stimulates the cooperative hydrolysis of all 7 ATPs yielding a stable GroEL–7ADP–GroES complex. Unfolded protein then binds to the toroid opposite the bound GroES, stimulating the release of the GroES and its rebinding, together with ATP, to either GroEL toroid. However, only protein bound to the same toroid as the GroES, the cis toroid, is stimulated to fold. The cooperative hydrolysis of the ATP in the cis toroid liberates the enclosed protein from its binding site(s) on GroEL, thereby permitting the protein to fold. The subsequent hydrolysis of ATP bound to the trans (opposite) toroid induces the departure of GroES together with the protein. However, a protein that is incompletely or incorrectly folded rebinds tightly to the same or another GroEL thereby initiating another round of folding. This process repeats until all the protein has properly folded. Note that, in contrast to Fig. 8-11*a*, the protein never passes between the GroEL toroids. The rate of hydrolysis of the trans toroid-bound ATP, which has a half-life of ~15 s, apparently acts as a timer for the folding process.

Chen, S., Roseman, A.M., Hunter, A.S., Wood, S.P., Burston, S.G., Ransom, N.A., Clarke, A.R., and Saibil, H.R., Location of a folding protein and shape changes in GroEL–GroES complexes imaged by cryo-electron microsopy, *Nature* **371,** 261–264 (1994).

Hayer-Hartl, M.K., Martin, J., and Hartl, F.U., Asymmetrical interaction of GroEL and GroES in the ATPase cycle of assisted protein folding, *Science* **269,** 836–841 (1995).

Mayhew, M., da Silva, A.C.R., Martin, J., Erdjument-Bromage, H., Tempst, P., and Hartl, F.U., Protein folding in the central cavity of the GroEL–GroES chaperonin complex, *Nature* **379,** 420-426 (1996).

Weissman, J.S., Rye, H.S., Fenton, W.A., Beechem, J.M., and Horwich, A.L., Characterization of the active intermediate of a GroEL–GroES-mediated protein folding reaction, *Cell* **84,** 481–490 (1996).

(d) LINUS: A Hierarchic Protein Folding Algorithm

Remarkable progress in solving the protein folding problem has been made through the computer program LINUS (which is both an acronym for "*L*ocal *I*ndependently *N*ucleated *U*nits of *S*tructure" and a tribute to Linus Pauling). The program is based on the observation that proteins are hierarchically organized such that small segments of contiguous polypeptide chain locally interact to form primitive folding modules, which associate with other such primitive folding modules to form larger modules, etc., thereby forming supersecondary structures, domains, and ultimately entire protein monomers. LINUS therefore simulates protein folding through a hierarchic condensation process in which local chain segments interact to form larger modules which, in turn, interact to form yet larger modules, etc.

In LINUS, a polypeptide of defined sequence is represented by idealized geometry and its interactions are governed by a highly simplified energy function consisting of four rules: (1) Conformations in which nonbonded atoms occupy the same space are forbidden (the hard sphere approximation); (2) a hydrogen bond receives one energy point; (3) interactions between two hydrophobic residues receive two energy points, those between a hydrophobic and an amphiphilic residue receive one point, and all other interactions receive zero points; and (4) residues with conformations on the right side of the Ramachandran diagram (Fig. 7-7), that is, with $\phi > 0°$, are penalized one point.

A folding simulation run begins with a fully extended polypeptide. Progressing one residue at a time from N- to C-terminus, three consecutive residues are conformationally perturbed at random yielding a trial conformation whose "energy" is then evaluated over an interaction interval of six residues. If the "energy" satisfies certain criteria, the trial conformation is kept; otherwise the previous conformation is kept. Several thousand cycles of this process are carried out in which a cycle is defined as a complete progression from N- to C-terminus. Persistent conformations are then kept and the entire process is iterated with progressively larger interaction intervals to a limit of 50 residues. Thus, large persisting segments, ranging from helices and β sheets to superecondary structures, are hierarchically built up as a concatenation of consecutively smaller segments.

This conceptually simple but computationally intensive algorithm has yielded surprisingly accurate structure predictions. In six of seven proteins of known X-ray structure whose folding LINUS has simulated, the secondary and supersecondary structures and large segments of the tertiary structure have been correctly albeit imprecisely predicted.

Srinivasan, R. and Rose, G.D., LINUS: A Hierarchic procedure to predict the fold of a protein, *Proteins* **22,** 81–99 (1995).

(e) Additional References

Darby, N.J., Morin, P.E., Talbo, G., and Creighton, T.E., Refolding of bovine pancreatic trypsin inhibitor via non-native disulfide intermediates, *J. Mol. Biol.* **249,** 463–467 (1995).

Weissman, J.S., All roads lead to Rome? The multiple pathways of protein folding, *Chem. Biol.* **2,** 255–260 (1996).

Dill, K.A., Bromberg, S., Yue, K., Fiebig, K.M., Yee, D.P., Thomas, P.D., and Chan, H.S., Principles of protein folding – A perspective from simple exact models, *Protein Science* **4,** 561–602 (1995).

Eisenhaber, F., Persson, B., and Argos, P., Protein structure prediction: Recognition of primary, secondary, and tertiary structural features from amino acid sequence, *Crit. Rev. Biochem. Mol. Biol.* **30,** 1–94 (1995).

Englander, S.W., Sosnick, T.R., Englander, J.J., and Mayne, L., Mechanisms and uses of hydrogen exchange, *Curr. Opin. Struct. Biol.* **6,** 18–23 (1996).

Fink, A.L., Compact intermediate states in protein folding, *Annu. Rev. Biophys. Biomol. Struct.* **24,** 495–522 (1995).

Honig, B. and Yang, A.-S., Free energy balance in protein folding, *Adv. Prot. Chem.* **46,** 27–58 (1995).

Karplus, M. and Sali, A., Theoretical studies of protein folding and unfolding, *Curr. Opin. Struct. Biol.* **5,** 58–73 (1995).

Lemer, C.M.-R., Rooman, M.J., and Wodak, S.J., Protein prediction by threading methods: Evaluation of current techniques, *Proteins* **23,** 337–355 (1995). [Threading is a structure prediction technique in which the known sequence of a protein of unknown structure is fitted to the known structures of proteins whose sequences have no obvious similarity to that of the protein in question.]

Miranker, A.D., and Dobson, C.M., Collapse and cooperativity in protein folding, *Curr. Opin. Struct. Biol.* **6,** 31–42 (1996).

Pain, R.H. (Ed.), *Mechanisms of Protein Folding,* IRL Press (1994).

Ptitsyn, O.B., Structures of folding intermediates, *Curr. Opin. Struct. Biol.* **5,** 74–78 (1995).

Wolynes, P.G., Onuchic, J.N., and Thirumalai, D., Navigating the folding routes, *Science* **267,** 1619–1620 (1996).

3. STRUCTURAL EVOLUTION

Aronson, H.-E.G., Royer, W.E., Jr., and Hendrickson, W.A., Quantification of tertiary structural conservation despite primary sequence drift in the globin fold, *Protein Science* **3,** 1706–1711 (1994).

Raine, A.R.C., Scrutton, N.S., and Mathews, F.S., On the evolution of alternate core packing in eighfold β/α-barrels, *Protein Science* **3,** 1889–1892 (1994).

Voet, D. and Voet, J.G., *KINEMAGES to Accompany Biochemistry, 2/E,* Wiley (1996). [Exercise 4 consists of kinemages on tuna cytochrome *c* and *Paracoccus* cytochrome c_{550}.]

Chapter 9

HEMOGLOBIN: PROTEIN FUNCTION IN MICROCOSM

2. STRUCTURE AND MECHANISM

Schlichting, I, Berendzen, J., Phillips, G.N., Jr., and Sweet, R.M., Crystal structure of photolysed carbonmonoxy-myoglobin, *Nature* **371,** 808–812 (1994). [An X-ray study performed at 20 K, which reveals the conformational changes that occur in carbonmonoxy-Mb upon the photodissociation of its covalently bound CO. This system has been used to study the time course of these conformational changes because they can be simultaneously triggered in all CO–Mb molecules by a laser pulse.]

Jayaraman, V., Rodgers, K.R., Mukerji, I., and Spiro, T.G., Hemoglobin allostery: resonance Raman spectroscopy of kinetic intermediates, *Science* **269,** 1843–1851 (1995). [A study of the time course of carbonmonoxy-hemoglobin's allosteric changes following the photodissociation of its bound CO in which the interactions of the heme group and specific Trp and Tyr residues are traced.]

Voet, D. and Voet, J.G., *KINEMAGES to Accompany Biochemistry, 2/E,* Wiley (1996). [Exercise 5 consists of kinemages of myoglobin, oxyhemoglobin, and deoxyhemoglobin together with comparisions of their tertiary and quaternary structures.]

3. ABNORMAL HEMOGLOBINS

(a) Hydroxyurea Therapy for Sickle-Cell Anemia

The treatment of individuals with sickle-cell anemia by the administration of hydroxyurea has recently been shown to ameliorate the symptoms of this painful and debilitating molecular disease. Hence hydroxyurea therapy is the first effective treatment that is specific for this inherited condition. Adults with sickle-cell anemia have two types of red blood cells: S cells, which contain only hemoglobin S (HbS); and F cells, which contain ~20% fetal hemoglobin (HbF) and the remainder HbS. In most adults, the fraction of F cells is ~30%. However, in those treated with hydroxyurea, this fraction increases to ~50%. Although the mechanism by which hydroxyurea stimulates the production of F cells is unknown, the mechanism by which increased levels of F cells prevent sickling seems clear. F cells contain three species of hemoglobin: HbS ($\alpha_2\beta^S_2$), HbF ($\alpha_2\gamma_2$), and their hybrid ($\alpha_2\beta^S\gamma$), where β^S subunits are the sickle-cell variants of the normal β subunits (Glu $\beta6 \rightarrow$ Val) and γ subunits are the β-like subunits of HbF. Since neither HbF nor the $\alpha_2\beta^S\gamma$ hybrid Hb can form sickle-cell fibers, they act to dilute the HbS in a cell. This, in turn, increases the time it takes the F cells to sickle by a factor of ~1000 (recall that the delay time for sickling, t_d, is an extremely sensitive function of the HbS concentration). This delay time is so long that F cells do not significantly sickle in the period (10-20 s) it takes them pass from the tissues to the lungs where they are oxygenated (recall that deoxyHbS but not oxyHbS forms sickle-cell fibers).Thus, the greater the proportion of F cells in the blood, the smaller the proportion of S cells that can sickle.

Charache, S., et al., Effect of hydroxyurea on the frequency of painful crises in sickle cell anemia, *N. Engl. J. Med.* **332,** 1317–1322 (1995).

Eaton, W.A. and Hofrichter, J., The biophysics of sicle cell hydroxyurea therapy, *Science* **268,** 1142–1143 (1995).

Chapter 10

SUGARS AND POLYSACCHARIDES

2. POLYSACCHARIDES

Voet, D. and Voet, J.G., *KINEMAGES to Accompany Biochemistry, 2/E,* Wiley (1996). [Exercise 6 contains kinemages on saccharides.].]

3. GLYCOPROTEINS

(a) Structure and Function of Heparin–bFGF Complexes

Protein growth factors act by binding with high affinity to their corresponding cell-surface receptors in a way that causes these transmembrane proteins to dimerize. This dimerization, in turn, generates a signal inside the the cell that initiates the growth factor's physiological response (Section 34-4B). A growing list of growth factors also bind with lower affinity to proteoglycans in a way that, by itself, does not generate a signal but modulates the biological response elicited when the growth factor binds to the receptor. For example, basic fibroblast growth factor (bFGF) must be in complex with the highly sulfated glycosaminoglycan heparin (Fig. 10-20) or the closely similar heparan sulfate in order to activate the FGF receptor.

The X-ray structures of bFGF in its complexes with both a tetrasacchararide and a hexasaccharide segment of heparin showed that both oligosaccharides bind to the same region of the bFGF molecules such that the tetrasaccharide closely superimposes on the four residues at the nonreducing end of the hexasaccharide. The heparin oligosaccharides assume helical structures that turn 174° and rise 8.6 Å per disaccharide unit, quanitities similar to the 180° and 8.0 to 8.7-Å values derived from X-ray fiber diffraction studies of heparin itself. The regions of bFGF to which the hexasaccharides bind is similar to those to which two heparin-derived trisaccharides bind in the X-ray structures of their complexes with bFGF. Curiously, despite the fact that neither of these trisaccharides is sulfated, their biological activities in facilitating bFGF activity are similar to that of heparin itself (recall that heparin is so highly sulfated that it is the most anionic polymer in mammalian tissues). Indeed, the addition of a sulfate group to one of these trisaccharides greatly reduces its biological activity.

The main chain conformations of the bFGF molcules in all of these complexes are essentially identical with that of bFGF alone. Thus, it is unlikely that heparin binding causes a conformational change in bFGF that modulates signal transduction. Rather, it has been proposed that heparin facilitates receptor dimerization either by binding multiple bFGF molecules or by binding both bFGF and its receptor.

Faham, S., Hileman, R.E., Fromm, J.R., Linhardt, R.J., and Rees, D.C., Heparin structure and interactions with basic fibroblast growth factor, *Science* **271,** 1116–1120 (1996).

(b) Additional References

Woods, R.J., Three-dimensional structures of oligosaccharides, *Curr. Opin. Struct. Biol.* **5,** 591–598 (1995).

Toone, E.J., Structure and energetics of proteins–carbohydrate complexes, *Curr. Opin. Struct. Biol.* **4,** 719–??? (1994).

Lee, Y.C. and Lee, R.T., Carbohydrate–protein interactions: Basis of glycobiology, *Acc. Chem. Res.* **28,** 321–327 (1995).

Chapter 11

LIPIDS AND MEMBRANES

2. PROPERTIES OF LIPID AGGREGATES

Lasic, D.D. and Papahadjopoulos, Liposomes revisited, *Science* **267,** 1275–1276 (1995). [Describes improvements to liposomes as drug-delivery vehicles.]

Pastor, R.W., Molecular dynamics and Monte Carlo simulations of lipid bilayers, *Curr. Opin. Struct. Biol.* **4,** 486–492 (1994). [Contains a computationally generated snapshot of the molecules in a lipid bilayer and their interacting water molecules.]

3. BIOLOGICAL MEMBRANES

(a) Detergent Structure in Crystals of OmpF Porin

The X-ray structure of OmpF porin in a new crystal form that has tetragonal symmetry has been determined. The molecular structure of OmpF in this tetragonal crystal is essentially identical to that in the previously determined trigonal crystal form (Fig. 10-28). In the trigonal crystals of OmpF, the hydrophobic surfaces of these transmembrane proteins are in contact, but in the tetragonal crystals, these surfaces are fully exposed. As the different intermolecular contacts in the two crystals do not significantly perturb the protein structure, it is very likely that the structure of OmpF in its native transmembrane environment closely resembles that in these crystals.

Since the membrane-exposed surfaces of integral membrane proteins such as OmpF are highly hydrophobic, they strongly and nonspecifically aggregate in aqueous solution. Such proteins must therefore be solubilized in a solution containing a detergent of sufficient potency to disrupt these hydrophobic protein–protein interactions but not so potent as to denature the protein. Thus, the solutions used to solubilize integral membrane proteins and from which they have been crystallized contain detergents with neutral rather than charged head groups (e.g., see Fig. 11-20).

In the tetragonal crystals of OmpF porin, the hydrophobic surfaces of the trimeric protein that contact lipids in the membrane are exposed and hence must interact with detergent. The nature of these protein–amphiphile interactions, a model for protein–lipid interactions, was investigated through single crystal neutron diffraction studies of tetragonal crystals of OmpF. [The detergent molecules are disordered in the crystal and hence are not seen as individual molecules. However, since deuterium nuclei (^{2}H) scatter neutrons quite differently from protons (^{1}H), detergent-containing regions can be distinguished from those containing only water by using deuterated detergent molecules.] The hydrophobic tails of the detergent molecules are localized in a ~25-Å high belt about the OmpF trimers which is located between the protein's surface-exposed bands of aromatic side chains (Fig. 11-28*c*). This observation supports the previously made hypothesis that these bands delimit those portions of OmpF that pass through the hydrophobic region of the lipid bilayer. Similar detergent belts have been observed in the crystal structures of two photosynthetic reaction centers, which also have bands of aromatic residues delimiting their detergent-binding regions.

Cowan, S.W., et al., The structure of OmpF porin in a tetragonal crystal form, *Structure* **3,** 1041–1050 (1995).

Pebay-Peyroula, E., Garavito, R.M., Rosenbusch, J.P., Zulauf, M., and Timmons, P.A., Detergent structure in tetragonal crystals of OmpF porin, *Structure* **3,** 1051–1059 (1995).

(b) Additional References

Kumar, N.M. and Gilula, N.B., The gap junction communication channel, *Cell* **84,** 381–388 (1996).

Reithmeier, R.A.F., Characterization and modeling of membrane proteins using sequence analysis, *Curr. Opin. Struct. Biol.* **5,** 491–500 (1995).

Shai, Y., Molecular recognition between membrane-spanning polypeptides, *Trends Biochem. Sci.* **20,** 460–464 (1995).

Stowell, M.H.B. and Rees, D.C., Structure and stability of membrane proteins, *Adv. Prot. Chem.* **46,** 279–311 (1995).

Voet, D. and Voet, J.G., *KINEMAGES to Accompany Biochemistry, 2/E,* Wiley (1996). [Exercise 7 consists of kinemages on bacteriorhodopsin, the photosynthetic reaction center from *Rps. viridis,* and OmpF porin.]

4. MEMBRANE ASSEMBLY AND PROTEIN TARGETING

(a) Of COPs, ARFs, SNAPs, and SNAREs: Mechanisms of Intracellular Protein Transport

Proteins are transferred between various membrane-enveloped compartments in the cell via small (500-1500-Å diameter) membranous vesicles. Vesicles that mediate endocytosis and transport from the trans Golgi network (Section 21-3B) to lysosomes and the plasma membrane have ~180-Å thick clathrin coats (Fig. 11-45), of which two types are known. At least two other types of vesicles are known to participate in membrane trafficking:(1) 700 to 750-Å diameter vesicles bearing a ~100-Å thick coat of **COPI proteins** (originally just COP, for *coat* protein), which were initially thought to mediate transport from the endoplasmic reticulum (ER) to the Golgi, but more recently have been implicated in the retrieval from the Golgi of ER-resident proteins that have the C-terminal sequence KKXX and transport between Golgi compartments; and (2) the more recently discovered 600 to 650-Å diameter vesicles bearing a ~100-Å thick coat of **COPII proteins,** which are the primary if not the exclusive mediators of ER-to-Golgi transport. COPI- and COPII-coated vesicles have a closely similar fuzzy appearance appearance under the electron microscope rather than the polyhedral shell of clathrin-coated vesicles (Fig. 11-45).

How are membranous vesicles formed, and how do they recognize and fuse with their target membranes? The formation of COPI-containing vesicles begins with the binding of an **ADP-ribosylation factor (ARF)** to a membrane. ARFs (which were first described as a cofactor in the cholera toxin-catalyzed ADP-ribosylation of G-protein α subunits; Section 34-4B) comprise a family of myristoylated proteins (Section 11-5A) that bind GTP and GDP. They are water-soluble cytosolic proteins in their GDP-bound form but in their GTP-bound form bind to membranes in a myristic acid-dependent manner. There are 7 COPI proteins, designated α-, β-, β', γ-,δ-, ϵ-, and ζ-COPs, that combine in one copy each to form a ~585 kD water-soluble complex named **coatamer.** Cytosolic coatamers bind to membrane-bound ARF–GTP complexes to form a coated bud, an all but complete coated vesicle that is not yet pinched off at its base. Since membranes remain flat on binding only ARF–GTP, it appears that ARF–GTP initiates budding by providing binding sites for coatamer, whereas coatamer binding to membrane-bound ARF–GTP mechanically drives budding by forming a curved array. Nevertheless, even when an excess of ARF, GTP, and coatamer is present, the buds do not pinch off. Rather, this requires the additional presence of a long-chain fatty acyl-CoA such as palmitoyl-CoA and is blocked by the presence of a nonhydrolyzable analog of palmitoyl-CoA.

Shortly after a COPI-coated vesicle pinches of from its parent membrane, it uncoats. This occurs through the ARF-mediated hydrolysis of its bound GTP to yield ARF–GDP, which no longer binds to the membrane and thereby also releases coatamer. Consequently, uncoating is inhibited by nonhydrolyzable analogs of GTP such as GTPγS. ARF by itself lacks GTPase activity. Rather, as with most of the numerous other members of the GTP-binding protein family (Section 30-3D), ARF requires the assistance of a GTPase activating protein (GAP) to hydrolzyze its bound GTP, and subsequently, of a guanine nucleotide releasing factor (GRF) to facilitate the exchange of the resulting GDP with GTP.

The uncoated vesicles are transported to their target membranes. In most cases, this occurs by simple diffusion, a process that typically takes from one to several minutes. Only in cases in which the vesicle must be transported large distances across the cytoplasm to reach its target membrane is this process facilitated by movement along cytoskeletal fibers (Section 34-3F).

Upon arriving at its target membrane, the vesicle fuses with it, thereby releasing its contents inside the target membrane (Fig. 11-46). This fusion process is blocked by low concentrations of the cysteine-alkylating agent *N*-ethylmaleimide (**NEM;** Table 6-3), indicating the presence of an **NEM-sensitive fusion (NSF) protein.** NSF is a cytosolic ATPase that does not bind to Golgi membranes unless **soluble NSF attachment proteins (SNAPs)** are also present. There are three types of SNAPs known: α- and γ-SNAPs are present in all cells so far examined and β-SNAP occurs only in brain. SNAPs bind to Golgi membrane in the absence of NSF indicating that SNAPs bind before NSF. α- and β-SNAPs bind competitively to the same sites on alkali-extracted membranes, which indicates that **SNAP receptors (SNAREs)** are integral membrane proteins.

Why do vesicles fuse with only their target membranes and not other membranes? A variety of evidence strongly suggests that each vesicle contains a unique address marker, a so-called **v-SNARE** (v for *v*esicle), which is obtained from its parent membrane during budding and which specifically binds to a so-called **t-SNARE** in its *t*arget membrane. Different vesicles have different v-SNARES, which are each targeted to a cognate t-SNARE in a specific membrane. SNAPs then assemble on the v-SNARE–t-SNARE complex followed by NSF, and a target membrane-specific **Rab protein**. The Rab proteins are a family of GTPases homologous to Ras protein (Section 34-4B) that are thought to function to verify the fit between a v-SNARE and its cognate t-SNARE: If the v-SNARE and t-SNARE remain bound together long enough for Rab to hydrolyze its bound GTP, this locks the vesicle to the target membrane. Membrane fusion then occurs via an ATP-dependent but otherwise poorly understood mechanism. It is postulated that the presence of coatamers prevents the v-SNARE-containing parent membrane from directly fusing with the t-SNARE-containing target membrane, and hence that vesicle uncoating is prerequisite for its fusion with its target membrane. This **SNARE hypothesis** may well describe a general mechanism of membrane fusion: It may apply not only to COPI-coated vesicles but also to COPII- and clathrin-coated vesicles and possibly to other types of vesicles yet to be characterized.

Amor, J.C., Harrison, D.H., Kahn, R.A., and Ringe, D., Structure of the human ADP-ribosylation factor 1 complexed with GDP, *Nature* **372,** 704–708 (1994).

Kreis, T.E., Lowe, M., and Pepperkok, R., COPs regulating membrane traffic, *Annu. Rev. Cell. Biol.* **11,** 677–706 (1995).

Mallabiabarrena, A. and Malhotra, V., Vesicle biogenesis: The coat connection, *Cell* **83,** 667–669 (1995).

Mellman, I., Enigma variations: protein mediators of membrane fusion, *Cell* **82,** 869–872 (1995).

Moss, J. and Vaughan, M., Structure and function of ARF proteins: Activators of cholera toxin and critical components of intracellular vesicular transport processes, *J. Biol. Chem.* **270,** 12327–12330 (1995).

Rothman, J.E., Mechanisms of intracellular protein transport, *Nature* **372,** 55–63 (1994).

Rothman, J.E. and Orci, L., Budding vesicles in living cells, *Sci. Amer.* **274** (3): 70–75 (1996).

Rothman, J.E. and Wieland, F.T., Protein sorting by transport vesicles, *Science* **272,** 227–234 (1996).

Schekman, R. and Orci, L., Coat proteins and vesicle budding, *Science* **271,** 1526–1533 (1996).

(b) Mitochondrial Protein Import Receptors

Studies using the common bread mold (*Neurospora crassa*) and Baker's yeast (*S. cerevisae*) have identified many of the components of the mitochondrial protein import system (Fig. 11-47) and clarified their functions. **Mitochondrial stimulation factor (MSF),** a heterodimer of 30 and 32 kD subunits, is a cytosolic ATP-dependent chaperone protein that specifically recognizes the N-terminal mitochondrial import signal on a precursor protein, unfolds it, and conducts it to a receptor on the mitochondrial outer membrane surface. This receptor consists of two loosely associated subcomplexes, the **MAS37–MAS70** and the **MAS20–MAS22** heterodimers. MAS37–MAS70 appears to interact with precursor proteins that have been unfolded by MSF and Hsp70, possibly acting in concert, and then transfers these proteins to MAS20–MAS22. Those proteins that do not require the ATP-dependent assistance of MSF or Hsp70 to unfold appear to bind directly to MAS20–MAS22. The unfolded proteins are then thought to cross the mitochondrial outer membrane via a passive pore or channel named **GIP** that consists of at least five subunits and which associates tightly with MAS20–MAS22.

The translocation of a precursor protein across the inner mitochondrial membrane involves the participation of a heterotetrameric core complex (**Mim14, Mim17, Mim23, and Mim30;** Mim for *m*itochondrial *i*nner *m*embrane) that, while a protein is being imported, interacts with a complex of **Mim44** and the matrix-resident chaperone protein mHsp70. It is thought that this heterotetrameric complex acts as a passive channel, whereas mHsp70, an ATPase, is the "motor" that drives the import process. The outer membrane (OM) and inner membrane (IM) protein translocation systems cooperate in translocating proteins. Thus, although isolated IM vesicles can fully translocate proteins into their lumens, isolated OM vesicles cannot complete this process. The manner in which the OM and IM protein import systems interact is unknown.

Lithgow, T., Glick, B.S., and Schatz, G., The protein import receptor of mitochondria, *Trends Biochem. Sci.* **20,** 98–101 (1995).

Ryan, K.R. and Jensen, R.E., Protein translocation across mitochondrial membranes: What a long, strange trip it is, *Cell* **83,** 517–519 (1995).

(c) Additional References

Hicks, G.R. and Raikhel, N.V., Protein import into the nucleus: An integrated view, *Annu. Rev. Cell Biol.* **11,** 155–188 (1995).

Schatz, G. and Dobberstein, B., Common principles of protein translocation across membranes, *Science* **271,** 1519–1526 (1996) [A wide ranging review.]

Siegel, V., A second signal recognition event required for translocation into the endoplasmic reticulum, *Cell* **82,** 167–170 (1995).

5. LIPOPROTEINS

(a) ApoE4 Is Implicated in Cardiovascular Disease and Alzheimer's Disease

Human apolipoprotein E (apoE), a 299-residue monomeric protein, is a component of all lipoproteins but LDL (Table 11-6). It consists of an N-terminal domain (residues 1-191), which is largely comprised of a 4-helix bundle and is thought to contain the protein's LDL receptor binding site (ApoE, as does apoB-100, specifically binds to the LDL receptor), and a C-terminal domain (residues 216-299), which binds to the lipoprotein surface.

There are three common alleles of apoE: **apoE2,** which has Cys at its positions 112 and 158; **ApoE3** (the most common isoform), in which these residues are Cys and Arg, respectively, and whose structure is shown in Fig. 11-58; and **apoE4,** in which these residues are both Arg. These differences have medical significance: ApoE3 has a preference for binding to HDL, whereas apoE4 has a preference for binding to VLDL, which is probably why apoE4 is associated with elevated plasma concentrations of LDL and thus an increased risk of cardiovascular disease. Evidently, changes in apoE's N-terminal domain can affect the function of its C-terminal lipoprotein-binding domain.

ApoE4 is also associated with a greatly increased incidence of **Alzheimer's disease (AD),** an ultimately fatal degenerative disease of the brain characterized by progressive loss of cognitive function. This observation is perhaps less surprising when it is realized that apoE is expressed by certain nerve cells and is present in the cerebrospinal fluid, where it apparently functions in mediating cholesterol transport, much as is does in blood plasma (cholesterol is abundant in nerve cell plasma membranes, which mediate neurotransmission; Section 34-4C).

Microscopic examination of brain tissue from AD victims reveals numerous extracellular **amyloid plaques,** which consist of fibrillar deposits of **amyloid β (Aβ) peptide** that apparently arise through proteolysis of the normally occurring **amyloid precursor protein.** It has been hypothesized that amyloid plaques are AD's pathogenic agent. Immunochemical staining indicates that apoE is associated with amyloid plaques. *In vitro* experiments demonstrate that both apoE3 and apoE4 form SDS-stable complexes with Aβ peptide that, after long incubation times, aggregate and precipitate from solution as a matrix of fibrils that closely resemble those in amyloid plaques. ApoE4 forms this complex more readily than apoE3 and yields a denser, more extensive matrix.

Brain tissue from AD victims also contains intracellular **neurofibrillary tangle**s, which consist mainly of paired helical filaments of hyperphosphorylated **tau protein.** ApoE3 forms an SDS-stable 1:1 complex with tau protein, whereas apoE4 forms little if any such complex. However, the phosphorylation of tau protein eliminates its ability to form a complex with apoE3. Tau protein facilitates the assembly and stability of microtubules and hence of the cytoskeleton (Section 34-3F). It has therefore been postulated that apoE3 but not apoE4 inhibits the phosphorylation of tau protein and thereby prevents microtubule instability and tangle formation.

Comparison of the X-ray structures of apoE4 and apoE3 reveals that there are only minor differences in their backbone conformations, which are restricted to the immediate vicinity of their site of difference (residue 112: Cys in ApoE3 and Arg in ApoE4). The only two side chains in apoE4 that undergo changes in conformation relative to those in apoE3 are Glu 109, which swings around to form a salt bridge with Arg 112, and Arg 61, which overlies Cys 112 in apoE3 but swings away to accommodate the new salt bridge in apoE4. Thus, both Glu 109 and Arg 61 are candidates for mediating the functional differences between apoE3 and apoE4. However, the mutagenic substitution of Ala for Glu 109 in apoE3 does not significantly alter its preference for binding to HDL over VLDL. In contrast, the substitution of Thr for Arg 61 in apoE4 gives this protein an apoE3-like preference for HDL over VLDL. Evidently, the position of Arg 61 is critical in determining the HDL/VLDL preference of apoE. This hypothesisis supported by the observation that residue 61 is invariably Thr in the 10 apoE's of known sequences from other species. None of these species exhibit the complete pathology of AD, although it remains to be demonstrated that Arg 61 actually contributes to the differential binding of apoE3 and apoE4 to Aβ peptide and tau protein.

Dong, L.-M., Wilson, C., Wardell, M.R., Simmons, T., Mahley, R.W., Weisgraber, K.H., and Agard, D.A., Human apolipoprotein E: Role of arginine 61 in mediating the lipoprotein preferences of the E3 and E4 isoforms, *J. Biol. Chem.* **269,** 22358–22365 (1994).

Weisgraber, K.H., Pitas, R.E., and Mahley, R.W., Lipoproteins, neurobiology, and Alzheimer's disease: structure and function of apolipoprotein E, *Curr. Opin. Struct. Biol.* **4,** 507–515 (1994).

Weisgraber, K.H., Apolipoprotein E: structure–function relationships, *Adv. Prot. Chem.* **45,** 249–301 (1995).

Strittmatter, W.J., Weisgraber, K.H., Huang, D.Y., Dong, L.-M., Salvesen. G.S., Pericak-Vance, M., Schmechel, D., Saunders, A.M., Goldgaber, D., and Roses, A.D., Binding of human apolipoprotein E to synthetic amyloid β peptide: Isoform-specific effects and implications for late-onset Alzheimer disease, *Proc. Natl. Acad. Sci.* **90,** 8098–8102 (1993).

(b) Additonal References

Acton, A., Rigotti, A., Landschulz, K.T., Xu, S., Hobbs, H.H., and Krieger, M., Identification of scavenger receptor SR-BI as a high density lipoprotein receptor, *Science* **271,** 518–520 (1996). [The first known HDL receptor, which mediates cholesterol uptake by a mechanism that differs from that of the LDL receptor pathway.]

Daly, N.L., Scanlon, M.J., Djordjevic, J.T., Kroon, P.A., and Smith, R., Three-dimensional structure of a cysteine-rich repeat from the low-density lipoprotein receptor, *Proc. Natl. Acad. Sci.* 92, 6334–6338 (1995).

Schumaker, V.N., Phillips, M.L., and Chatterton, J.E., Apolipoprotein B and low-density lipoprotein structure: Implications for biosynthesis of triglyceride-rich lipoproteins, *Adv. Prot. Chem.* **45,** 205–248 (1995).

Tall, A., Plasma lipid transfer proteins, *Annu. Rev. Biochem.* **64,** 235–267 (1995). [Discusses the properties and physiology of cholesteryl ester transfer protein (CETP).]

Chapter 12

INTRODUCTION TO ENZYMES

2. Substrate Specificity

Lamzin, V.S., Dauter, Z., and Wilson, K.S., How nature deals with stereoisomers, *Curr. Opin. Struct. Biol.* **5,** 830–836 (1995).

Ringe, D., What makes a binding site a binding site? *Curr. Opin. Struct. Biol.* **5,** 825–829 (1995).

4. REGULATION OF ENZYMATIC ACTIVITY

Koshland, D.E., Jr., The key–lock theory and induced fit theory, *Angew. Chem. Int. Ed. Engl.* **33,** 2375–2378 (1994). [Recounts the history of the induced fit theory in the context of Fischer's lock-and-key hypothesis of enzyme activity.]

Voet, D. and Voet, J.G., *KINEMAGES to Accompany Biochemistry, 2/E,* Wiley (1996). [Exercise 8 consists of kinemages on R- and T-state apartate transcarbamoylases and their comparison.]

Chapter 13

RATES OF ENZYMATIC REACTIONS

2. ENZYME KINETICS

Gutfreund, H., *Kinetics for the Life Sciences: Receptors, Transmitters, and Catalysts,* Cambridge (1995)

Schulz, A.R., *Enzyme Kinetics,* Cambridge (1994).

Chapter 14

ENZYMATIC CATALYSIS

1. CATALYTIC MECHANISMS

Gerlt, J.A., Protein engineering to study enzyme catalytic mechanisms, *Curr, Opin. Struct. Biol.* **4,** 593–600 (1994). [Descriptions of case studies of enzyme mechanisms using protein engineering.]

Kyte, J., *Mechanism in Protein Chemistry,* Garland (1995).

Moffat, K. and Henderson, R., Freeze trapping of reaction intermediates, *Curr, Opin. Struct. Biol.* **5,** 656–663 (1993). [Discusses techniques for structurally characterizing the intermediates of enzymatic reactions at low temperatures and the results of such studies.]

2. LYSOZYME

(a) More on the Mechanism of Lysozyme

Despite nearly 35 years of mechanistic and structural investigations on hen egg white lysozyme (HEWL), two aspects of the Phillips mechanism are still disputed: (1) the existence of enzyme-induced D-ring distortion; and (2) whether the reaction proceeds via a double displacement mechanism rather than via the single displacement envisioned in the Phillips mechanism. In addition, no X-ray structure yet determined has shown saccharide units in HEWL's E and F subsites. The explanation for this is that native HEWL rapidly hydrolyzes any oligosaccharide spanning its D and E subsites and releases the E and F-bound product.

In an effort to visualize an entire hexasaccharide bound to HEWL, the X-ray structure of the HEWL mutant Asp 52 → Ser (D52S HEWL) crystallized in the presence of $(NAG)_6$ was determined. Although Asp 52 is an important catalytic residue (recall that the Phillips mechanism postulates that it functions to electrostatically stabilize the reaction's oxonium ion intermediate), D52S HEWL exhibits a residual catalytic activity that is <1% of that of native HEWL [recall that the Asp 52 → Asn (D52N) mutant exhibits ~5% activity]. Thus, over the several day crystallization period, it hydrolyzed the $(NAG)_6$ to $(NAG)_4$ + $(NAG)_2$. Consequently, the X-ray structure exhibited only the presence of $(NAG)_4$, which was bound to HEWL's subsites A through D. The NAG bound in the D subsite clearly assumed a half-chair conformation with its atom C6 in an axial position (as was similarly observed for the D subsite-bound NAM in the complex of NAM-NAG-NAM with HEWL), in accordance with the Phillips mechanism.

An unexpected observation is that the D ring-bound NAG in the D52S HEWL complex has the α configuration about its anomeric carbon atom (C1) despite the observation that native HEWL hydrolyzes β(1 → 4) glycosidic linkages with 99.9% retention of configuration. Comparison of the D52S and native HEWL structures reveals a space in the "α pocket" of D52S HEWL that is filled by the larger Asp 52 side chain in native HEWL. This supports Phillips' conjecture that retention of configuration in the HEWL reaction arises through the steric exclusion of water from approaching C1 from the α direction.

An alternative to the Phillips mechanism postulates that the carboxyl group of Asp 52 forms a covalent bond to C1 of the oxonium ion, thereby yielding a covalent glycosyl–enzyme intermediate that is subsequently hydrolyzed to yield product. Such a double displacement mechanism (as opposed to the Phillips single displacement mechanism), which has been clearly demonstrated to occur in several types of glycosidases, would account for the observed retention of configuration in the HEWL reaction. However, it is at odds with the observation that the distance between C1 in a D subsite-bound saccharide and a carboxyl O of Asp 52 (which participates in a network of hydrogen bonds that hold this side chain in its position) are far too long to form a covalent bond. Indeed, no such covalent bond has been observed in any HEWL-containing X-ray structure. Further evidence disfavoring the double displacemeent mechanism has been provided by the structure of goose egg white lysozyme (GEWL) in complex with $(NAG)_3$. GEWL has little sequence identity with HEWL but these two proteins nevertheless share common tertiary structural elements. The $(NAG)_3$ occupies subsites analogous to the B, C, and D subsites of HEWL and Glu 73 of GEWL corresponds closely to Glu 35 of HEWL. However, GEWL has no obvious counterpart to HEWL's Asp 52 and hence GEWL cannot carry out a double displacement reaction. This is also the case for the D52S and D52N HEWLs, which afterall, are both good catalysts in that they mediate rate enhancements of ~10^6 over the uncatalyzed rate of substrate hydrolysis. Thus, a second carboxyl group is not essential to the catalytic activity of GEWL and may not be necessary for the activities of other lysozymes.

Hatfield, A.T., Harvey, D.J., Archer, D.B., MacKenzie, D.A., Jeenes, D.J., Radford, S.E., Lowe, G., Dobson, C.M., and Johnson, L.N., Crystal structure of the mutant D52S hen egg white lysozyme with an oligosaccharide product, *J. Mol. Biol.* **243,** 856–872 (1994).

Kirby, A.J., Turning lysozyme upside down, *Nature Struct. Biol.* **2,** 923–925 (1995); *and* Illuminating an ancient retainer, *Nature Struct. Biol.* **3,** 107–108 (1996). [Argues, based on organic mechanistic principles as well as the known structures and chemistry of certain glycosidases, that the reaction catalyzed by HEWL is likely to proceed via a double displacement mechanism.]

Weaver, L.H., Grütter, M.G., and Matthews, B.W., The refined structures of goose lysozyme and its complex with a bound trisaccharide show that the "goose-type" lysozymes lack a catalytic aspartate residue, *J. Mol. Biol.* **245,** 54–68 (1995).

(b) Additional References

Davies, G. and Henrissat, B., Structures and mechanisms of glycosyl hydrolases, *Structure* **3,** 853–859 (1995).

McCarter, J.D. and Withers, S.G., Mechanisms of enzymatic glycoside hydrolysis, *Curr. Opin. Struct. Biol.* **4,** 885–892 (1994).

Voet, D. and Voet, J.G., *KINEMAGES to Accompany Biochemistry, 2/E,* Wiley (1996). [Exercise 9 contains a kinemage on HEW lysozyme.]

3. SERINE PROTEASES

(a) Structural Basis of Substrate Specificity in Trypsin and Chymotrypsin

Trypsin catalyzes the hydrolysis of peptidyl amide substrates with an Arg or Lys residue preceding the scissile bond with an efficiency, as measured by k_{cat}/K_M (Section 13-2B), that is 10^6-fold greater than that for the corresponding Phe-containing substrates. Conversely, chymotrypsin catalyzes the hydrolysis of substrates after Phe, Trp, and Tyr residues 10^4-fold more efficiently than after the corresponding Lys-containing substrates. The structural basis of this selectivity is more complicated than simply replacing the negatively charged Asp 189, which lies at the bottom of trypsin's specificity pocket (in which Lys and Arg side chains bind; Fig. 14-20), with chymotrypsin's corresponding Ser residue. Indeed, doing so does not yield a chymotrypsin-like enzyme but, rather, a nonspecific protease with greatly reduced catalytic efficiency. Only when all the residues of trypsin's specificity pocket and those of two surface loops flanking the active site, L1 (residues 185-188) and L2 (residues 221-225), are replaced with the corresponding residues of chymotrypsin, does the resulting enzyme (termed Tr→Ch[S1+L1+L2]), have a chymotrypsin-like specificity and acylation rate, although it still has a low substrate-binding affinity, K_S. However, the additional mutation Tyr 172 → Trp (Y172W) in a third surface loop, yields an enzyme (Tr→Ch[S1+L1+L2+Y172W]) with 15% of chymotrypsin's catalytic efficiency.

Careful comparisons of the X-ray structures of chymotrypsin and trypsin with those of the closely similar Tr→Ch[S1+L1+L2], and Tr→Ch[S1+L1+L2+Y172W] in complex with a Phe-containing chloromethyl ketone inhibitor reveal the structural basis of substrate specificity in trypsin and chymotrypsin. Efficient catalysis in the serine proteases requires that the enzyme's active site be structurally intact and that the substrate's scissile bond be properly positioned relative to the catalytic triad and the oxyanion hole. The above mutagenic changes do not affect the structure of the catalytic triad or those portions of the active site that bind the substrate's leaving group (that segment on the C-terminal side of the scissile bond). However, the main chain conformation of the conserved Gly 216 (which forms two hydrogen bonds to the backbone of the third residue before the substrate's scissile bond in an antiparallel β pleated sheet-like arrangement) differs in trypsin and chymotrypsin and adopts a chymotrypsin-like structure in both hybrid proteins. Evidently, if Gly 216 adopts a trypsin-like conformation, the scissile bond in Phe-containing substrates is misoriented for proper catalysis. Thus, despite the fact that Gly 216 is conserved in trypsin and chymotrypsin, the differing structures of loop L2 in the two enzymes maintains it in distinct conformations.

Loop L1, which interacts with L2 in both trypsin and chymotrypsin, is largely disordered in Tr→Ch[S1+L1+L2]. Modeling a trypsin-like L1 into Tr→Ch[S1+L1+L2] results in severe steric clashes with the chymotrypsin-like L2. Thus, the requirement of a chymotrypsin-like L1 for the efficient catalysis by Tr→Ch[S1+L1+L2] appears to arise from the need to permit L2 to adopt a chymotrypsin-like conformation.

Residue 172 is located at the base of the specificity pocket. The improvement in substrate binding affinity of Tr→Ch[S1+L1+L2+Y172W] over Tr→Ch[S1+L1+L2] arises from structural rearrangements in this region of the enzyme caused by the increased bulk and different hydrogen bonding requirements of Trp vs Tyr. These changes appear to improve both the structural stability of residues forming the specificity pocket and their specificity for chymotrypsin-like substrates.

Perona, J.J., Hedstrom, L., Rutter, W.J., and Fletterick, R.J., Structural origins of substrate discrimination in trypsin and chymotrypsin, *Biochem.* **34,** 1489–1499 (1995).

Perona, J.J. and Craik, C.S., Structural basis of substrate specificity in the serine proteases, *Prot. Sci.* **4,** 337–360 (1995). [An authoritative review.]

(b) The Direct Observation of an Acyl-Enzyme Intermediate in Elastase

Although there is a great variety of evidence (e.g., Fig. 14-28) favoring the transient existence of an acyl-enzyme intermediate in the reaction mechanism of serine proteases (Fig. 14-23), such an intermediate in a productive reaction pathway had heretofore never been directly observed at atomic resolution (that reported on *pp.* 395-396 of the textbook had less-than-atomic resolution and hence the acyl group was only visualized as a bilobal but otherwise featureless blob). The 2.3-Å resolution X-ray structure of this intermediate in the elastase-catalyzed hydrolysis of the slowly deacylating substrate *N*-carbobenzoxy-L-Ala-*p*-nitrophenyl ester (ZAP) has now been observed. This acyl-enzyme intermediate was formed by soaking a crystal of the enzyme in a ZAP-containing solution (recall that protein crystals contain labyrinthine solvent-filled passages through which small solute molecules can readily pass). This reaction was carried out at –26°C, a temperature at which the intermediate's already slow deacylation rate is greatly reduced such that the majority of the enzyme molecules assumed their acylated form (the soaking solution in this process contained 70% methanol as an antifreeze). The temperature of the crystal was then rapidly reduced to –55°C, below the glass transition temperature at which the collective motions an enzyme requires to catalyze a reaction are arrested (Section 8-2), thereby preserving this acyl-enzyme intermediate. The X-ray structure of this crystal revealed that the acyl-enzyme intermediate has its expected planar structure.

Ding, X., Rasmussen, B.F., Petsko, G.A., and Ringe, D., Direct structural observation of an acyl-enzyme intermediate in the hydrolysis of an ester substrate by elastase, *Biochemistry* **33,**

9285–9293 (1994).

(c) Additional References

Roberts, R.M., Mathialagan, N., Duffy, J.Y., and Smith, G.W., Regulation and regulatory role of proteinase inhibitors, *Crit. Rev. Euk. Gene. Express.* **5,** 385–436 (1995).

Voet, D. and Voet, J.G., *KINEMAGES to Accompany Biochemistry, 2/E,* Wiley (1996). [Exercise 10 consists of kinemages on chymotrypsin, trypsin, subtilisin and their enzymatic mechanisms.]

Warshel, A., Papazyan, A., and Kollman, P.A.; Cleland, W.W. and Kreevoy, M.M.; *and* Frey, P.A., On low-barrier hydrogen bonds and enzyme catalysis; *and* Responses *Science* **269,** 102–106 (1995). [Theoretical arguments that the catalytic effects which have been attributed to low-barrier hydrogen bonds are better explained by electrostatic and solvation effects. Both responses reject these notions and reassert the catalytic significance of low-barrier hydrogen bonds.]

4. GLUTATHIONE REDUCTASE

Voet, D. and Voet, J.G., *KINEMAGES to Accompany Biochemistry, 2/E,* Wiley (1996). [Exercise 11 consists of kinemages on glutathione reductase and its catalytic cycle.]

Chapter 16

GLYCOLYSIS

2. THE REACTIONS OF GLYCOLYSIS

(a) The Flexible Loop of TIM Is Not Ligand-Gated

Triose phosphate isomerase (TIM) has a flexible loop that closes over its substrate so as to help bind it to the enzyme (Fig. 16-11), thereby excluding bulk solvent from the active site and possibly modifying its dielectric constant. Moreover, the flexible loop holds the enediol(ate) intermediate's phosphate group in the plane of the enediol(ate), a conformation that stereoelectronically prevents the formation of the methyl glyoxal side product, which would otherwise dominate the reaction (Fig. 16-12).

Comparison of the X-ray structures of TIM with and without bound substrate suggests that the motion of the flexible loop is ligand-gated, that is, the binding of substrate induces the loop to change conformation such that it closes down over the substrate. However, solid-state deuterium NMR measurements indicate that, when TIM binds the substrate analog glycerol-3-phosphate or the transition state analog 2-phosphoglycolate, the loop moves at a rate similar to that of the unbound enzyme and has a similar population ratio for its two conformers. These measurements further indicate that the loop opens and closes on the 100-μs time scale, which is slightly faster than the enzyme's turnover time (the single conformations of the loop observed in TIM-containing crystal structures are apparently artifacts of crystallization).

It had been widely assumed that loop closure in TIM is ligand-gated. Yet, if this were the case, consideration of the reversibilty of the TIM reaction and the chemical resemblance of its reactant and product (GAP and DHAP) makes it difficult to rationalize how product could be released. However, a rate of spontaneous loop opening and closing similar to TIM's catalytic turnover time would permit facile product release but only infrequent loop opening while the intermediate was bound to the enzyme.

Williams, J.C. and McDermott, A.E., Dynamics of the flexible loop of triosephosphate isomerase: The loop motion is not ligand-gated, *Biochemistry* **34,** 8309–8319 (1995).

Voet, D. and Voet, J.G., *KINEMAGES to Accompany Biochemistry, 2/E,* Wiley (1996). [Exercise 12 consists of kinemages on the structure and mechanism of TIM.]

(b) Additional References

Mattevi, A., Valentini, G., Rizzi, M., Speranza, M.L., Bolognesi, M., and Coda, A., Crystal structure of *Eschericia coli* pyruvate kinase type I: molecular basis of allosteric transition, *Structure* **3,** 729–741 (1995).

Gefflaut, T., Blonski, C., Perie, J., and Willson, M., Class I aldolases: Substrate specificity, mechanism, inhibitors and structural aspects, *Prog. Biophys. Molec. Biol.* **63,** 301–340 (1995).

3. FERMENTATION: THE ANAEROBIC FATE OF PYRUVATE

Lesk, A.M., NAD-binding domains of dehydrogenases, *Curr. Opin. Struct. Biol.* **5,** 775–783 (1995).

4. CONTROL OF METABOLIC FLUX

(a) Internal Motions in Adenylate Kinase Function as an Energetic Counterweight to Balance Substrate Binding

The adenylate kinase (AK) reaction (2ADP $\rightarrow$ ATP + AMP) rapidly equilibrates the ADP resulting from ATP hydrolysis in muscle with ATP and AMP, thus buffering the ATP concentration during glycolysis and muscle contraction.. AK must be specific to prevent undesirable phosphoryl transfer reactions such as hydrolysis. This specificity is accomplished by the closure of two ~30-residue domains over the substrates, thereby tightly binding them. Once the reaction has occurred, the tightly bound products must be rapidly released to maintain the enzyme's catalytic efficiency. Yet, since the AK reaction is energetically neutral (it replaces one phosphodiester bond with another), what is the origin of the energy necessary to rapidly release its products?

Comparison of the X-ray structures of unliganded AK with AK in complex with the inhibitory bisubstrate analog Ap$_5$A (2 ADPs connected by a fifth phosphate) suggests how AK avoids falling into the energy well of tight-binding substrates and products. Upon binding substrate, a portion of the protein remote from the active site increases its chain mobility and "resolidifies" on product release. Thus, the enzyme has a sort of internal binding energy that is released to help bind substrate and reabsorbed when product is released. This, it is hypothesized, acts as an "energetic counterweight" to permit facile product release and hence maintain a high reaction rate.

Müller, C.W., Schlauderer, G.J., Reinstein, J., and Schulz, G.E., Adenylate kinase motions during catalysis: an energetic counterweight balancing substrate binding, *Structure* **4,** 147–156 (1996).

(b) Additional References

Voet, D. and Voet, J.G., *KINEMAGES to Accompany Biochemistry, 2/E,* Wiley (1996). [Exercise 13 consists of kinemages on the structure and allosteric mechanism of PFK.]

Chapter 17

GLYCOGEN METABOLISM

3. CONTROL OF GLYCOGEN METABOLISM

(a) Structure of the Regulatory Subunit of cAMP-Dependent Protein Kinase (cAPK)

The phosphorylase kinase-catalyzed phosphorylation of glycogen phosphorylase's Ser 14 converts this enzyme from its allosterically responsive and mostly inactive *b* form to its allosterically largely unresponsive and mostly active *a* form (Fig. 17-13, *upper left*). Phosphorylase kinase is itself activated by phosphorylation in a reaction catalyzed cAMP-dependent protein kinase (cAPK). In the absence of cAMP, cAPK is a catalytically inactive heterotetramer that consists of two regulatory (R) subunits and two catalytic (C) subunits, R_2C_2. The R subunits each bind two cAMP molecules such that, in the presence of a sufficiently high concentration of cAMP, the C subunits (whose structure is shown in Fig. 17-14) dissociate as catalytically active monomers from resulting the $R_2(cAMP)_4$ dimer.

The R subunit has a well defined domain structure that was first characterized by limited proteolysis. It consists of, from N- to C-terminus, a dimerization domain, an autoinhibitor segment, and two tandem homologous cAMP-binding domains, A and B. In the R_2C_2 complex, the autoinhibitor segment, which resembles the C subunit's substrate, binds in the C subunit's active site (as does the inhibitory peptide in Fig. 17-14), thereby blocking substrate binding.

Each R subunit cooperatively binds 2 cAMPs. The B domain masks the A domain so as to prevent it from binding cAMP. However, the binding of cAMP to the B domain triggers a conformational change that permits the A domain to bind cAMP, which in turn, releases the C subunits from the complex.

The X-ray structure of the R subunit lacking its N-terminal 91 residues and in complex with two cAMP's has been determined. This truncated protein is unable to dimerize but, in the absence of cAMP, forms a tight inactive complex with the C subunit, and upon binding cAMP, releases active C subunits as do intact R_2 dimers. As previously predicted by sequence alignments, the A and B domains are structurally similar to each other and to the prokaryotic cAMP-binding protein named catabolite gene activator protein (CAP; whose X-ray structure is shown in Fig. 29-22a). Extensive interactions between the A and B domains presumably mediate the conformational change that "opens up" domain A for cAMP binding when cAMP has bound to the B domain.The autoinhibitor segment, which in the free R subunit is extremely sensitive to proteolysis, has its first 21 residues disordered in the X-ray structure.

Su, Y., Dostmann, W.R.G., Herberg, F.W., Durick, K., Xuong, N., Ten Eyck, L., Taylor, S.S., and Varughese, K.I., Regulatory subunit of protein kinase A: Structure of deletion mutant with cAMP binding domains, *Science* **269**, 807–813 (1995).

(b) Structure of the Phosphorylase Kinase Catalytic Domain

Phosphorylase kinase (**Phk**) is a ~1300 kD protein of subunit composition $(\alpha\beta\gamma\delta)_4$. The γ subunit contains Phk's catalytic site, whereas its α, β, and δ subunits have regulatory functions. The δ subunit is calmodulin (CaM), which in the presence of Ca^{2+}, activates the γ subunit. However, maximum activity is conferred only if, in addition, the α and β subunits are phosphorylated. The protein kinases known to phosphorylate the α and β subunits of Phk include cAMP-dependent protein kinase (cAPK), AMP-dependent protein kinase (AMPK; Section 23-4B), and Phk itself.

Phk's 386-residue γ subunit contains of an N-terminal kinase domain, which is 36% identical in sequence to cAPK, and a C-terminal regulatory domain, which contains CaM-binding regions including a putative pseudosubstrate segment. The pseudosubstrate segment is thought to regulate kinase activity by binding to and thereby blocking the kinase active site (as does the inhibitory peptide in the cAPK structure; Fig. 17-14). However, when Ca^{2+} is present, the pseudosubstrate binds to the resulting Ca^{2+}-activated CaM subunit (as does the MLCK autoinhibitory peptide in Fig. 17-17), thereby freeing the kinase active site. Thus, the N-terminal 298-residue segment of the Phk γ subunit, termed **Phkγtrnc**, displays catalytic activity comparable to that of fully activated Phk but is unaffected by Ca^{2+} or phosphorylation signals.

The X-ray structures of Phkγtrnc in complex with adenosine-5'-(β,γ-imido)triphosphate (AMP-PNP, *p.* 589) + Mn^{2+} and with ADP + Mg^{2+} have been determined. As expected, the Phkγtrnc in both complexes structurally resembles cAPK (Fig. 17-14) as well as other protein kinases of known structure [CDK2 (Fig. 33-79), MAP kinase (Section 34-4B), **CaMKI**, and **twitchin kinase** (see Section17-3d below)].

Comparisons of these various structures sheds light on how the catalytic activity of Phk is regulated. Numerous protein kinases, including CDK2 and MAP kinase, are activated by phosphorylation of one or two Ser, Thr and/or Tyr residues in their so-called activation loop, which is located at the "mouth" of the cleft between the kinase's N- and C-terminal domains. The added phosphate group(s) form a network of interactions that stabilize the kinase's catalytic machinery in an active conformation. In particular, all protein kinases that are activated by phosphorylation have a conserved Arg residue that is adjacent to a catalytically implicated Asp residue and which is positioned to interact with the activating phosphate group. The Phk γ subunit is not subject to phosphorylation. Rather, the residue that might otherwise be phosphorylated is Glu, whose negative charge, it is thought, mimics the presence of a phosphate group and is positioned such that it interacts with the conserved Arg residue. Thus, the Phk catalytic site maintains an active conformation but, in the absence of Ca^{2+}, is inactivated by the binding of the pseudosubstrate segment. The way in which the phosphorylation state of the α and β subunits modulates the kinase activity is, as yet, unkown.

Owen, D.J., Noble, M.E.M., Garman, E.F., Papageorgiou, A.C., and Johnson, L.N., *Structure* **3**, 467–482 (1995).

(c) The Nature of Calmodulin's Calcium-Induced Conformational Changes

Calmodulin (CaM) in complex with Ca^{2+} binds to and thereby activates its target proteins, many of which are protein kinases that, in turn, activate or deactivate their target proteins. Thus, CaM functions to couple a transient Ca^{2+} influx caused by a stimulus to a cell-surface receptor to intracellular events (Fig. 34-108). What structural changes does Ca^{2+} binding impose on CaM that permit it to act as a Ca^{2+}-activated switch?

Four independent back-to-back papers on the solution structures of calmodulin (CaM) and the closely related troponin C (TnC; Fig. 34-58) in the absence of calcium ion, and the comparisons of these structures to those of the corresponding Ca^{2+} complexes, reveal the nature of these Ca^{2+}-induced conformational changes. Ca^{2+} binding does not alter CaM's secondary structure. However, Ca^{2+} induces similar tertiary structural changes in all four of CaM's Ca^{2+}-binding EF-hand motifs, thereby exposing an otherwise buried Met-rich hydrophobic patch on the surface of each of CaM's two domains to which CaM's target peptides bind (Fig. 17-17). These studies also reveal that the polypeptide segment (residues 77-80) which links CaMs two globular domains is flexible in solution rather than assuming a helical conformation as it does in the crystal structure (Fig. 17-15). This flexible linker is nevertheless essential to CaM's functions: In the presence of Ca^{2+}, CaM's individual domains (obtained by tryptic cleavage), when in high concentration, are able to bind to some of their target proteins but, nevertheless, fail to even marginally activate them unless present in several hundred-fold excess.

Zhang, M., Tanaka, T., Ikura, M, Calcium-induced conformational transition revealed by the solution structure of apo calmodulin, *Nature Struct. Biol.* **2**, 758–767 (1995).

Koboniwa, H., Tjandra, N., Grzesiek, S., Ren, H., Klee, C.B., and Bax, A., Solution structure of calcium-free calmodulin, *Nature Struct. Biol.* **2**, 768–776 (1995).

Finn, B.E., Evenäs, J., Drakenberg, T., Waltho, J.P., Thulin, E., and Forsén, S., Calcium-induced structural changes and domain autonomy in calmodulin, *Nature Struct. Biol.* **2**, 777–783 (1995).

Gagné, S.M., Tsuda, S., Li, M.X., Smillie, L.B., and Sykes, B.D., Structures of troponin C regulatory domains in the apo and calcium-saturated states, *Nature Struct. Biol.* **2**, 784–789 (1995).

James, P., Vorherr, T., and Carafoli, E., Calmodulin-binding domains: just two faced or multi-faceted? *Trends Biochem. Sci.* **20**, 38–42 (1995).

(d) The Structural Basis of Autoregulation in Protein Kinases

Protein kinases constitute a large family of enzymes whose catalytic activity is tightly coupled to a variety of extracellular and intracellular signals. Many protein kinases are inactive in their "resting" state but are activated by the binding of a specific regulator protein, such as Ca^{2+}-CaM, phosphorylation by certain protein kinases, or both. In the prevailing theory of the phosphorylation-independent activation of protein kinases, the inactive protein kinase binds a non-phosphorylatable pseudosubstrate that mimics the kinase's polypeptide target (e.g., Fig. 17-14), thereby inactivating the kinase. However, a Ca^{2+}-activated protein such as CaM (see Section 17-3c above) specifically binds to the pseudosubstrate so as to extract it from the kinase active site, yielding active kinase.

The X-ray structures of two homologous protein kinases support this so-called intrasteric mechanism, those of **calcium/calmodulin-dependent protein kinase I (CaMKI)** and **twitchin kinase.** CaMKI is a monomeric 374-residue enzyme that occurs in many tissues and can phosphorylate a number of proteins. Twitchin kinase is a 373-residue segment of **twitchin,** an enormous (6839 residue) protein that is situated in the muscle A-band (Fig. 34-68) of the nematode *Caenorhabditis elegans* and so-named because mutant twitchins confer impaired movements including a "twitch" of the body surface. Both CaMKI and twitchin kinase consist of an N-terminal protein kinase domain and a C-terminal pseudosubstrate tail that binds to the kinase active site so as to inhibit it. The 51-residue C-terminal tail of CaMKI contains a segment to which Ca^{2+}-CaM binds, whereas twitchin kinase has no obvious Ca^{2+}-CaM-binding segment and hence it is unclear how this enzyme is activated.

The N-terminal kinase portions of CaMKI and twitchin kinase both form bilobal structures that resemble those of cAPK (Fig. 17-14) and other protein kinases of known structure (see Section 17-3b above). The C-terminal regulatory segment of CaMKI forms a helix–loop–helix motif that extends across both domains of the catalytic center. The N-terminal helix and the loop interfere with the binding of polypeptide substrates, whereas the loop and the C-terminal helix interact with the periphery of the ATP-binding site so as to induce conformational changes that obstruct the binding of ATP and ADP. Since the structure of CaMKI reveals that its Trp 303, a component of its Ca^{2+}-CaM recognition segment, is exposed, it is proposed that Ca^{2+}-CaM extracts the pseudosubstrate from CaMKI's active site by initially binding to Trp 303. The C-terminal regulatory segment of twitchin kinase also forms a helix–loop–helix motif that interacts with its kinase active site in a manner that resembles that in CaMKI but is different in detail. Thus, both structures provide direct support for the intrasteric mechanism of kinase activation.

Goldberg, J., Nairn, A.C., and Kuriyan, J., Structural basis for the autoinhibition of calcium/calmodulin-dependent protein kinase I, *Cell* **84,** 875–887 (1996).

Hu, S.-H., Parker, M.W., Lei, J.Y., Wilce, M.C.J., Benian, G.M., and Kemp. B.E., Insights into autoregulation from the crystal structure of twitchin kinase, *Nature* **369,** 581–584 (1994).

(e) Additional References

Barford, D. and Johnson, L.N., Electrostatic effects in the control of glycogen phosphorylase by phosphorylation, *Protein Sci.* **3,** 1776–1730 (1994).

Crivici, A. and Ikura, M., Molecular and structural basis of target recognition by calmodulin,

Annu. Rev. Biophys. Biomol. Struct. **25,** 85–116 (1995).

Goldsmith, E.J. and Cobb, M.H., Protein kinases, *Curr. Opin Struct. Biol.* **4,** 833–840 (1994); *and* Cox, S., Radzio-Andzelm, E., and Taylor, S.S., Domain movements in protein kinases, *Curr. Opin Struct. Biol.* **4,** 893–901 (1994).

Pilkis, S.J., 6-Phosphofructo-2-kinase/fructose-2,6-bisphosphatase: a metabolic signaling enzyme, *Annu. Rev. Biochem.* **64,** 799–835 (1995).

Pilkis, S.J., Weber, I.T., Harrison, R.W., and Bell, G.I., Glucokinase: Structural analysis of a protein involved in susceptibility to diabetes, *J. Biol. Chem.* **269,** 21925–21928 (1994).

Voet, D. and Voet, J.G., *KINEMAGES to Accompany Biochemistry, 2/E,* Wiley (1996). [Exercise 15 is a kinemage of the catalytic subunit of cAPK in complex with its.target peptide and ATP. Exercise 16 consists of kinemages of Ca^{2+}_4–CaM, both alone and in complex with its target peptide.]

Chapter 18

TRANSPORT THROUGH MEMBRANES

2. KINETICS AND MECHANISM OF TRANSPORT

(a) The Structure of Maltoporin Reveals a Mechanism for Sugar Transport

Maltoporin is a homotrimeric transmembrane protein that facilitates the diffusion of **maltodextrins** [the $\alpha(1 \rightarrow 4)$-linked glucose oligosaccharide degradation products of starch; e.g., maltose (Fig. 10-12)] across the outer membrane of gram-negative bacteria. The X-ray structure of maltoporin from *E. coli* (the product of its *lamB* gene, which was first discovered as the receptor for bacteriophage λ; Section 32-3) reveals that maltoprin is structurally similar to OmpF porin (Figs. 11-28 and 18-13), but with an 18-stranded rather than a 16-stranded antiparallel β barrel enclosing each subunit's transport channel. Three long loops from the extracellular face of each maltoporin subunit fold inwards into the barrel, thereby constricting the channel near the center of the membrane to a diameter of 5 or 6 Å (which is significantly smaller than the 7×11-Å aperture in OmpF porin) and hence giving the channel an hourglass-like cross section. The channel is lined with a series of 6 contiguous aromatic side chains arranged in a left-handed helical path that matches the left-hand helical curvature of α-amylose (Fig. 10-17). This so-called "greasy slide" extends from the channel's vestibule floor, through its constriction, to its periplasmic outlet.

The way that maltodextrins interact with maltoporin was investigated by determining the X-ray structures of maltoporin in its complexes with the maltooligosaccharides Glc_2 (maltose),

Glc_3, and Glc_6. Two Glc_2 molecules, one Glc_3 molecule, and a Glc_5 unit of Glc_6 were seen to occupy the maltoporin channel in contact and conformity with the greasy slide. Many of the glucose residue hydroxyl groups are hydrogen bonded to polar side chains that line the opposite side of the channel. Six of these seven polar side chains are charged, which would probably strengthen their hydrogen bonds as has been observed in other complexes of sugars with proteins.

On the basis of the above structures and the observation that the hydrophobic faces of glycosyl residues stack on aromatic side chains in sugar-binding proteins, the following model for the selective transport of maltodextrins by maltoporin has been proposed. At the start of the translocation process, the entering glucosyl residue would interact with the readily accessible end of the greasy slide in the vestibule of the channel. Further translocation along the helical channel requires the maltodextrin to follow a screw-like path that maintains the helical structure of the oligosaccharide, much like the movement of a bolt through a nut, thereby excluding molecules of comparable size that have different shapes. The translocation process is unlikely to encounter any large energy barriers due to the smooth surface of the greasy slide and the multiple polar groups at the channel constriction that would permit the essentially continuous exchange of hydrogen bonds. Thus, maltoporin can be regarded as an enzyme that catalyzes the translocation of its substrate from one compartment to another.

Schirmer, T., Keller, T.A., Wang, Y.-F., and Rosenbusch, J.P., Structural basis for sugar translocation through maltoporin channels at 3.1 Å resolution, *Science* **267,** 512–514 (1995).

Dutzler, R., Wang, Y.-F., Rizkallah, P.J., Rosenbusch, J.P., and Schirmer, T., Crystal structure of various maltooligosaccharides bound to maltoporin reveal a specific sugar translocation pathway, *Structure* **4,** 127–134 (1996).

(b) Additional References

Herzberg, O. and Klevit, R., Unraveling a bacterial hexose transport pathway, *Curr. Opin. Struct. Biol.* **4,** 814–822 (1994).

Katz, E.B., Stenbit, A.E., Hatton, K., DePinho, R., and Charron, M.J., Cardiac and adipose tissue abnormalities but not diabetes in mice deficient in GLUT4, *Nature* **377,** 151–155 (1995). [Mice lacking a functional gene for GLUT4, an insulin-sensitive and the most abundant passive-mediated glucose transporter in muscle and adipose tissue, unexpectedly do not have greatly elevated blood glucose levels. Nevertheless, GLUT4 is essential for sustained growth, normal cellular glucose and fat metabolism, and expected longevity.]

Saier, M.H., Jr., Chauvaux, S., Deutscher, J., Reizer, J., and Ye, J.-J., Protein phosphorylation and regulation of carbon metabolism in Gram-negative versus Gram-positive bacteria, *Trends Biochem. Sci.* **20,** 267–271 (1995).

Voet, D. and Voet, J.G., *KINEMAGES to Accompany Biochemistry, 2/E,* Wiley (1996). [Exercise 7 contains a kinemage on OmpF porin.]

3. ATP-DRIVEN ACTIVE TRANSPORT

Lingrel, J.B. and Kuntzweiler, T., Na$^+$,K$^+$-ATPase, *J. Biol. Chem.* **269,** 19659–19662 (1994).

Chapter 19

CITRIC ACID CYCLE

2. METABOLIC SOURCES OF ACETYL-COENZYME A

(a) Intersubunit Interactions in the Pyruvate Dehydrogenase Multienzyme Complex

The pyruvate dehydrogenase multienzyme complex consists of (depending on the species) a cubic or a dodecahedral core of dihydrolipoamide transacetylase (E_2), to which multiple copies of pyruvate dehydrogenase (E_1) and dihydrolipoamide dehydrogenase (E_3) are noncovalently associated (Figs. 19-4 and 19-5). E_2 has a segmented structure that consists of three independently folded domains, which are covalently tethered by long highly flexible polypeptide linkers (Fig. 19-8). From N- to C-terminus, these domains are: (depending on the species) one, two, or three ~80-residue lipoyl domains, each of which contains a lipoic acid prosthetic group in Schiff base to a Lys side chain; a ~35-residue binding domain (E_2BD), one the smallest known globular domains; and a ~250 residue catalytic (acetyltransferase) domain, which forms the complex's cubic or dodecahedral core. E_2BD binds to E_3 in cubic complexes and to both E_1 and E_3 in dodecahedral complexes. The flexible linkers permit the lipoyl group to "visit" all three of the complex's active sites (to catalyze reactions 2, 3, and 4 in Fig. 19-6). E_3, a homodimer, is a member of the disulfide oxidoreductase family, which structurally and functionally resembles glutathione reductase (Section 14-4).

Both E_2BD of *Bacillus stearothermophilus* (which has a dodecahedral E_2 core) and the di-domain (E_2DD), which consists of E_2BD linked to the lipoyl domain, bind in 1:1 mole ratio to the E_3 homodimer. The X-ray structure of both these complexes are closely similar because the lipoyl domain appears to be fully disordered in the E_2DD–E_3 complex. The structure of E_3 in these complexes closely resembles that of E_3 alone. The structure of E_2BD in the complexes is very similar to that of E_2BD in solution as determined by NMR. Hence E_2BD binds to E_3 by more of a lock-and-key mechanism than an induced fit mechanism. The reason that only one E_2BD subunit binds to the E_3 dimer is that its binding site is so close to E_3's 2-fold axis of symmetry that a second symmetrically related E_2BD would sterically clash with the first. The structural knowledge gained in this study combined with previously determined structures of E_2's catalytic and lipoyl domains has lead to a plausible model of the entire E_2 in complex with E_3 that lets the lipoyl domain of E_2 swing around to "visit" the active sites of E_2 and E_3 in the multienzyme complex.

Mande, S.S., Sarfaty, S., Allen, M.D., Perham, R.N., and Hol, W.G.J., Protein–protein interactions in the pyruvate dehydrogenase multienzyme complex: dihydrolipoamide dehydrogenase complexed with the binding domain of dihydrolipoamide acetyltransferase, *Structure* **4,** 277–286 (1996).

(b) Additional References

Green, J.D.F., Laue, E.D., Perham, R.N., Ali, S.T., and Guest, J.R., Three-dimensional structure of a lipoyl domain from the dihydrolipoyl acetyltransferase component of the pyruvate dehydrogenase multienzyme complex of *Eschericia coli, J. Mol. Biol.* **248,** 328–343 (1995). [The NMR structure of an *E. coli* lipoyl domain reveals a sandwich of two 4-stranded antiparallel β sheets in which the lipoylation site occupies a physically exposed position in a tight turn connecting two of the β strands. The structure is closely similar to that of the previously determined lipoyl domain from the *B. stearothermophilus* enzyme.]

3. ENZYMES OF THE CITRIC ACID CYCLE

(a) Visualization of Intermediates in the Isocitrate Dehydrogenase Reaction

Isozymes of isocitrate dehydrogenase catalyze the NAD^+- or the $NADP^+$-dependent oxidative decarboxylation of isocitrate to α-ketoglutarate via an oxalosuccinate intermediate (Fig. 19-14), for whose existence there has been only indirect evidence. This is because this intermediate has but a highly transient existence in reactions catalyzed by the wild-type enzyme. However, an enzymatic reaction rate can be slowed by the mutation of particular catalytically important residues, leading to the accumulation of specific intermediates. Thus, the Y160F and the K230M mutations of isocitrate dehydrogenase create bottlenecks in the enzymatic reaction pathway that lead to the accumulation of the ternary isocitrate–Mg^{2+}–$NADP^+$ Michealis complex and the oxalosuccinate intermediate, respectively. These intermediates were directly visualized by the X-ray structure determinations of their corresponding mutant enzymes using fast (Laue) X-ray intensity measurement techniques so as to complete these measurements before the intermediate of interest could decay. Both of these mutagenized proteins have essentially the same structure as the wild-type enzyme.

Bolduc, J.M., Dyer, D.H., Scott, W.G., Singer, P., Sweet, R.M., Koshland, D.E., Jr., and Stoddard, B.L., Mutagenesis and Laue structures of enzyme intermediates: isocitrate dehydrogenase, *Science* **268**, 1312–1318 (1995).

(b) Additional References

Weaver, T.M., Levitt, D.G., Donnelly, M.I., Stevens, P.P.W., and Banaszak, L.J., The multisubunit active site of fumarase C from *Eschericia coli, Nature Struct. Biol.* **2,** 654–662 (1995). [The X-ray structure of a bacterial fumarase homologous to eukaryotic fumarases reveals that it is a homotetrameric enzyme whose subunits are related by three mutually perpendicular 2-fold axes (D_2 symmetry) and whose active sites are each comprised of residues from three of the four subunits.]

Chapter 20

ELECTRON TRANSPORT AND OXIDATIVE PHOSPHORYLATION

2. ELECTRON TRANSPORT

(a) X-Ray Structures of Cytochrome *c* Oxidase (Complex IV)

Cytochrome *c* oxidase, also known as complex IV, is the terminal enzyme of the respiratory electron transport chain. This transmembrane multisubunit protein mediates the transfer of electrons from 4 cytochrome *c* (Fe^{2+}) to O_2 yielding 4 cytochrome *c* (Fe^{3+}) and H_2O and harnessing the liberated free energy to translocate (pump) as many as four protons out of the mitochondrion or bacterial cell, thereby helping create a proton gradient whose dissipation powers the generation of ATP (Section 20-4). The protein complex contains two *a*-type cytochrome centers (Fig. 20-17), whose heme groups are named heme *a* and heme a_3, and two Cu centers, termed Cu_A and Cu_B. Spectroscopic studies indicate that electrons are passed from cytochrome *c* to the Cu_A center, then to the cytochrome *a*, and finally to a binuclear complex of cytochrome a_3 and the Cu_B center. O_2 binds to this binuclear complex and is reduced to H_2O in a complex 4-electron process (Fig. 20-22).

The X-ray structures of two species of cytochrome *c* oxidase were essentially simultaneously reported: A relatively simple form from the soil bacterium *Paracoccus denitrificans* and a more complicated form from bovine heart mitochondria. Cytochrome *c* oxidase from *P. denitrificans,* as viewed from within the plasma membrane in which it is embedded, has a trapezoidal outline that is 55 Å high, ~90 Å wide at its cytosolic surface, ~75 Å wide at its periplasmic surface, and is largely comprised of 22 mostly helical membrane-spanning segments and their connections. Subunit I, which contains 12 of these transmembrane segments, binds heme *a* and the heme a_3–Cu_B binuclear center. The heme a_3 Fe has one axial His ligand, the heme *a* Fe has two axial His ligands, and the Cu_B atom has three His ligands. There is, at odds with spectroscopic data, no bridging ligand evident between the heme a_3 Fe atom and Cu_B. The distance between the Fe atom of heme a_3 and Cu_B is 5.2 Å, the closest approach of the two heme molecules is 4.7 Å, and the distance between their Fe atoms is 13.2 Å. Subunit II contains two transmembrane segments and a globular domain on the periplasmic surface that binds the Cu_A center and largely consists of a 10-stranded β barrel similar to that of blue copper proteins such as plastocyanin (Fig. 22-20*b*). The Cu_A center, which for many years was widely thought to contain but one Cu atom, clearly contains two Cu atoms, which are bridged by two Cys S atom ligands and have two additional ligands each to form an arrangement similar to that of a [2Fe–2S] cluster (*p.* 574). Subunits III and IV, which have unknown functions, contain one and seven transmembrane helices, respectively. The cytochrome c binding site is postulated to be in a corner formed by the globular domain of subunit II and the periplasmic surface of subunit I since this region is close to the Cu_A site and contains 10 acidic side chains that could interact with the Lys side chains on the surface of cytochrome *c* (Fig. 20-19). The structure suggests possible pathways for electron transfer and proton translocation.

Cytochrome c oxidase from bovine heart mitochondia is a ~200-kD transmembrane protein that consists of 13 different subunits. The monomeric unit of this dimeric protein has an ellipsoidal (potato-like) shape rather than the tooth-like or crenellated shapes seen in Fig. 20-21. It has a 48-Å thick transmembrane segment and hydrophilic portions that protrude 32 and 37 Å into the mitochondrial matrix and the intermembrane space, respectively. The atomic details of only the metal ion-binding sites in this protein have yet been reported. Subunit I, which has 12 transmembrane helices, binds the heme a, heme a_3, and Cu_B centers, which are all located ~13 Å below the membrane surface on its cytosolic side. The Cu_A site, which also contains two Cu atoms, is bound to subunit II and is located 8 Å above the membrane surface on its cytosolic side. The relative geometric arrangements and liganding patterns of the redox-active metal centers are closely similar to those in the *P. denitrificans* enzyme. Subunits I, II, and III, which are all encoded by mitochondrial genes, form the core of the structure, while its other 11 subunits, which are all encoded by nuclear genes and whose functions are unknown, form a sheath around this core.

Iwata, S., Ostermeier, C., Ludwig, B., and Michel, H., Structure at 2.8 Å resolution of cytochrome c oxidase from *Paracoccus denitrificans, Nature* **376,** 660–669 (1995).

Tsukihara, T., Aoyama, H., Yamashita, E., Tomizaki, T., Yamaguchi, H., Shinzawa-Itoh, K., Nakashima, R., Yaono, R., and Yoshikawa, S., Structures of metal sites of oxidized bovine heart cytochrome c oxidase at 2.8 Å, *Science* **269,** 1069–1074 (1995).

Scott, R.A., Functional significance of cytochrome c oxidase structure, *Nature Struct. Biol.* **3,** 981–986 (1995). [Reviews the forgoing papers.]

(b) Additional References

Voet, D. and Voet, J.G., *KINEMAGES to Accompany Biochemistry, 2/E,* Wiley (1996). [Exercise 17 is a kinemage of the 1:1 complex of cytochrome c with cytochrome c peroxidase.]

Zhou, J.S., Nocek, J.M., DeVan, M.L., and Hoffman, B.M., Inhibitor-enhanced electron transfer: Copper cytochrome c as a redox-inert probe of ternary complexes, *Science* **269,** 204–207 (1995). [Demonstrates, through spectroscopic measurements, that cytochrome c peroxidase forms a ternary complex with two molecules of cytochrome c in which one cytochrome c molecule binds tightly to a surface domain of cytochrome c peroxidase that has a low electron transfer potential and the second cytochrome c molecule binds weakly to the 1:1 complex but with 1000-fold greater reactivity.]

3. OXIDATIVE PHOSPHORYLATION

Walker, J.E., The regulation of catalysis in ATP synthase, *Curr. Opin. Struct. Biol.* **4,** 912–918 (1994).

Wilkens, S., Dahlquist, F.W., McIntosh, L.P., Donaldson, L.W., and Capaldi, R.A., Structural features of the ε subunit of the *Eschericha coli* ATP synthase determined by NMR spectroscopy, *Nature Struct. Biol.* **2,** 961–967 (1995).

Chapter 21

OTHER PATHWAYS OF CARBOHYDRATE METABOLISM

3. BIOSYNTHESIS OF OLIGOSACCHARIDES AND GLYCOPROTEINS

Takeda, J., and Kinoshita, T., GPI-anchor biosynthesis, *Trends Biochem. Sci.* **20,** 367–371 (1995).

Udenfriend, S. and Kodukula, K., How glycosylphosphatidyinositol-anchored membrane proteins are made, *Annu. Rev. Biochem.* **64,** 563–591 (1995).

4. THE PENTOSE PHOSPHATE PATHWAY

(a) G6PD Deficiency Confers Resistance to Severe Malaria in Both Males and Females

The geographical distribution of glucose-6-phosphate dehydrogenase (G6PD) deficiency, the most common enzymopathy in humans, suggests that this X-linked inherited disease confers protection against malaria. Previous studies had indicated, as reported in the textbook, that only heterozygous females are afforded antimalarial protection; males (who are hemizygous) and homozygous females appeared to be unprotected. However, this phenomenon of differential protection has been disputed.

In two epidemiological studies involving over 2000 African children with the most common African form of G6PD deficiency, the A⁻ form, it was found that this form is associated with a ~50% reduction in risk of severe malaria for both female heterozygotes and male hemizygotes (female G6PD A⁻ homozygotes were too rare to permit the determination of their susceptibility to severe malaria). Female heterozygotes also had statistically significant protection against mild malaria but this was not the case with males. This degree of protection from malaria is less than that afforded to carriers of the sickle-cell trait (Sections 6-3A and 9-3B) but equal or greater than that associated with some thalassemias (Section 33-2G) and certain MHC variants (Section 34-2E).

Ruwende, et al., Natural selection of hemi- and heterozygotes for G6PD deficiency in Africa by resistance to severe malaria, *Nature* **376,** 246–249 (1995).

(b) Additional References

Beutler, E., G6PD deficiency, *Blood* **84,** 3613–3636 (1994). [A review.]

Chapter 22

PHOTOSYNTHESIS

2. LIGHT REACTIONS

(a) Light Harvesting Complexes in Purple Photosynthetic Bacteria

Most purple photosynthetic bacteria have two types of light-harvesting complexes, **LH1** and **LH2,** with different spectral and biochemical properties. LH2 rapidly passes the energy from the photons it absorbs to LH1 which, in turn, passes it to the photosynthetic reaction center. The structures of representatives of these transmembrane proteins have been determined.

LH2 from *Rhodopseudomonas acidophila* strain 10050 consists of 9 copies each of two types of subunits: its α-apoprotein (53 residues) and its β-apoprotein (41 residues), which collectively bind 27 bacteriochlorophyll *a* (BChl *a*) molecules, and 9 carotenoid molecules. The X-ray structure of LH2 reveals an elegant 9-fold rotationally symmetric arrangement. The α- and β-apoproteins both consist almost entirely of single helices and are aligned nearly perpendicularly to the plane of the membrane in which they are embedded. The 9 α-apoproteins pack side-by-side to form a hollow cylinder of radius 18 Å. Each of the 9 β-apoproteins occupies a position radially outwards from an α-apoprotein to form a concentric cylinder of radius 34 Å. 18 of the BChl *a* molecules are packed between these rings of helices in an arrangement resembling an 18-bladed turbine: Successive nearly parallel BChl *a* ring systems are in partial van der Waals contact with their planes perpendicular to the plane of the membrane and their centers all ~10 Å below the presumed periplasmic membrane surface. The Mg^{2+} atoms in these BChl *a* molecules are each singly axially liganded by His side chains that alternately extend from an α- and a β-apoprotein around the ring. The remaining 9 BChl *a* molecules, which are each singly axially liganded by the carbonyl oxygen atom of the α-apoprotein's N-terminal formyl-Met (fMet) residue, are arranged in a 9-fold symmetric ring between successive β-apoprotein helices and oriented with the planes of their ring systems parallel to the plane of the membrane and hence all in a single plane, which is located 16.5 Å below the ring of His-liganded BChl *a* molecules (11 Å from the membrane's cytoplasmic surface). The 9 carotenoid molecules are sandwiched between the α- and β-apoproteins and extend along much of their lengths thereby contacting both sets of BChl *a* molecules. The LH2 from *Rhodovulum sulfidophilum* has a similar structure as determined at low (7-Å) resolution by electron crystallography.

Spectrocopic measurements indicate that LH2's 18 His-liganded and closely associated BChl *a* molecules maximally absorb radiation at a wavelength of 850 nm (and hence are called B850) and are strongly coupled, that is, they absorb radiation almost as a unit. The 9 fMet-liganded BChl *a* molecules (B800) maximally absorb radiation at 800 nm, largely as individual molecules. When a B800 BChl *a* absorbs a photon, the excitation is rapidly (in 700 fs) transferred to a lower energy B850 BCl *a* (which may independently absorb a photon), which even more rapidly (in 200-300 fs) exchanges the excitation among the other B850 BCh *a* molecules. Hence the B850 system acts as a kind of energy storage ring that delocalizes the excitation over a large region. The carotenoids in this system absorb visible (<800 nm) light and may facilitate the transmission of excitation between the rather distantly separated (21 Å between Mg atoms) nearest neighbor B850 and B800 BChl *a* molecules.

LH1, like LH2, has α and β subunits of ~50 residues each. The low (8.5-Å) resolution structure of LH1 from *Rhodospirillum rubrum*, as determined by electron crystallography, reveals it to resemble LH2 but with 16-fold rotational symmetry and hence to form a 116-Å diameter cylinder with a 68-Å diameter hole down its center. This hole is of sufficient size to contain a photosynthetic reaction center (Figs. 11-27 and 22-9). LH1's BChl *a* molecules absorb radiation at a longer wave length than those of LH2 and consequently, when these two assemblies are in contact, excitation is rapidly (in 1-5 ps) transferred from LH2 to LH1 and then (in 20-40 ps) to LH1's enclosed photosynthetic reaction center. Excitations may also be rapidly exchanged between contacting LH2s. Thus, this antenna system transfers virtually all of the radiation energy it absorbs to the photosynthetic reaction center in far less than the few ns over which these excitations would otherwise decay. It should be noted that this complicated arrangement of chromophores is among the simplest known; those of the light-harvesting systems of plants are even more elaborate (e.g., Fig. 22-8*b*).

McDermott, G., Prince, S.M., Freer, A.A., Horthornthwaite-Lawless, A.M., Papiz, M.Z., Cogdell, R.J., and Isaacs, N.W., Crystal structure of an integral membrane light-harvesting complex from photosynthetic bacteria, *Nature* **374,** 517–521 (1995). [The X-ray structure of LH2].

Karrasch, S., Bullough, P.A., and Ghosh, R., The 8.5 Å projection map of the light harvesting complex I from *Rhodspirillum rubrum* reveals a ring composed of 16 subunits, *EMBO J.* **14,** 631–638 (1995).

Kühlbrandt, W., Structure and function of bacterial light-harvesting complexes, *Structure* **3,** 521–525 (1995); *and* Isaacs, N.W., Light-harvesting mechanisms in purple photosynthetic bacteria, *Curr. Opin. Struct. Biol.* **5,** 794–797 (1995). [Reviews.]

Savage, H., Cyrklaff, M., Montoya, G., Kühlbrandt, W., and Sinning, I., Two-dimensional structure of light-harvesting complex II (LHII) from the purple bacterium *Rhodovulum sulfidophilum* and comparison with LHII from *Rhodopseudomonas acidophila*, *Structure* **4,** 243–252 (1996).

(b) A Mutant Photosynthetic Reaction Center that Transfers Electrons to its M-Side Chromophores

In the photosynthetic reaction centers of the purple photosynthetic bacteria, the homologous L and M subunits together with their bound chromophores are arranged with approximate 2-fold symmetry. Yet, as Fig. 22-10 indicates, this assembly transfers electrons exclusively along its L-side chromophores with a quantum yield of 100%. Mutant forms of the photosynthetic reaction center from *Rhodobacter capsulatus* have provided insights into the origin of this enigmatic asymmetry.

Changing Leu M212 to His places this new His side chain over the face of the L-side BPheo *a* in an Mg^{2+}-liganding position, thereby converting it to a BChl *a* molecule, which is named β. The L(M212)H mutant transfers 65% of its charge-separated electrons along its L-side chromophores, with the remaining 35% undergoing rapid charge recombination and return to the ground state. This is because the higher free energy of electron transfer to β relative to that of the original BPheo *a* results in partial electron transfer from the light-excited special pair to the so-called accessory BChl a, which in the wild-type reaction center does not accept electrons.

Additionally mutating Gly M201 to Asp, which places this negatively charged side chain near ring V (Fig. 22-3) of the L-side accessory BChl *a*, raises the free energy of this chromophore to the point that it cannot accept electrons from the special pair. Hence, electron transfer through the L side of this G(M201)D/L(M212)H double mutant can proceed only through β. This further changes the system's photochemistry to 70% electron transfer to L-side chromophores (via β), 15% rapid charge recombination, and 15% electron transfer to the previously nonparticipating M-side BPheo *a*. Evidently, individual amino acid residues, through their influence on the free energies of charge-separated states, can facilitate electron transfer to the M-side chromophores.

Heller, B.A., Holten, D., and Kirmaier, C., Control of electron transfer between the L- and M-sides of photosynthetic reaction centers, *Science* **269,** 940–945 (1995).

(c) Photosystem II Can Generate O_2 Without the Aid of Photosytem I

A mutant form of *Chlamydomonas reinhardtii* that lacks photosystem I (PSI) can, nevertheless, photosynthetically assimilate CO_2 and generate O_2 and H_2 under anaerobic conditions, but not in air. Thus, under proper conditions, photosystem II (PSII) alone has sufficient light-induced redox potential to generate the NADPH necessary to reduce CO_2. Indeed, the quantum yields for NADPH production are similar for both the mutant and the wild-type *Chlamydomonas*. This suggests that the evolution of the water-splitting photosynthetic apparatus initially involved only PSII, which arose from a primordial bacterial photosynthetic reaction center. PSI, and hence the Z-scheme, only arose after the O_2 level in the atmosphere had increased to levels that impaired the function of PS2.

Greenbaum, E., Lee, J.W., Tevault, C.V., Blankinship, S.L., and Mets, L.J., CO_2 fixation and photoevolution of H_2 and O_2 in a mutant *Chlamydomonas* lacking photosystem I, *Nature* **376,** 438–441 (1995).

(d) Additional References

Arnoux, B., Gaucher, J.-F., Ducruix, A., and Reiss-Husson, F., Structure of the photochemical reaction centre of a spheroidene-containing purple bacterium, *Rhodobacter sphaeroides* Y, at 3 Å resolution, *Acta Cryst.* **D51,** 368–379 (1995). [Reports the structure of a photosynthetic reaction center closely similar to other known reaction center structures but in which the Fe^{2+} ion in these other reaction centers is replaced by a similarly liganded Mn^{2+} ion.].

Ermler, U., Fritzsch, G., Buchanan, S.K., and Michel, H., Structure of the photosynthetic reaction centre from *Rhodobacter sphaeroides* at 2.65 Å resolution: cofactors and protein–cofactor interactions, *Structure* **2,** 925–936 (1994); *and* Deisenhofer, J., Epp, O., Sinning, I., and Michel, H., Crystallographic refinement at 2.32 Å resolution and refined model of the photosynthetic reaction centre from *Rhodopseudomonas viridis*, *J. Mol. Biol.* **246,** 429–457 (1995). [Both structures reveal long strings of contacting water molecules leading from the protein-enclosed quinone Q_B to the cytoplasmic side of the protein and which may represent a proton transfer pathway to Q_B^{2-}].

Rögner, M., Boekema, E.J., and Barber, J., How does photosystem 2 split water? The structural basis of efficient energy conversion, *Trens Biochem. Sci.* 21, 44–49 (1996). [Presents an electron microscopy-based model for the subunit arrangement in PS2 and discusses its significance.]

Voet, D. and Voet, J.G., *KINEMAGES to Accompany Biochemistry, 2/E,* Wiley (1996). [Exercise 7 contains a kinemage on the photosynthetic reaction center from *Rps. viridis* showing its arrangement of subunits, chromophores, and its electron transfer pathway.]

Chapter 23

LIPID METABOLISM

1. LIPID DIGESTION, ABSORPTION, AND TRANSPORT

Egloff, M.-P., Marguet, F., Buono, G., Verger, R., Cambillau, C., and van Tillbeurgh, H., The 2.46 Å resolution structure of the pancreatic lipase–colipase complex inhibited by a C11 alkyl phosphonate, *Biochemistry* 34, 2751–2762 (1995). [The inhibitor is covalently linked to the lipase's active site Ser and its C_{11} alkyl chain is bound in a hydrophobic groove of the protein in a way that is thought to mimic the the interaction between the fatty acid leaving group of a triacylglycerol substrate with the protein.]

Gelb, M.H., Jain, M.K., Hanel, A.M., and Berg, O.G., Interfacial enzymology of glycerolipid hydrolases, *Annu, Rev. Biochem.* 64, 653–688 (1995).

Woolley, P. and Petersen, S.B. (Eds.), *Lipases,* Cambridge Univ. Press (1994)

2. FATTY ACID OXIDATION

(a) Structure of Methylmalonyl-CoA Mutase

Methylmalonyl-CoA mutase catalyzes the carbon skeleton rearrangement of (*R*)-methylmalonyl-CoA to succinyl-CoA with the assistance of a coenzyme B_{12} (adenosylcobalmin) prosthetic group. It is the only coenzyme B_{12}-containing enzyme that is present in both prokaryotes and eukaryotes.

Methylmolonyl-CoA from *Propionibacterium shermanii* is an αβ heterodimer whose catalytically active 728-residue α subunit is 24% identical in sequence to its catalytically inactive 638-residue β subunit. In contrast, the human enzyme is an α_2 homodimer whose sequence is 60% identical to that of the *P. shermanii* α subunit.

The X-ray structure of *P. shermanii* methylmalonyl-CoA mutase in complex with the partial substrate desufo-CoA (which lacks the substrate's S atom and methylmalonyl group) reveals that its subunits are structurally homologous with an N-terminal domain that consists of an α/β barrel (TIM barrel, the most common enzymatic motif ; Section 16-2E) and a C-terminal α/β domain that contains a core of 5 parallel β strands. Coenzyme B_{12} is largely contained within the catalytically active α subunit's C-terminal domain, which is packed against the bottom of the α/β barrel (at the N-terminal ends of the barrel's β strands). In free coenzyme B_{12}, the Co atom is axially liganded by the 5'-CH_2 group of its adenosyl residue and by an N atom of its 5,6-dimethylbenzimidazole group (Fig. 23-19). However, in the enzyme, neither of these ligands are present. Rather, the 5,6-dimethylbenzimidazole group has swung aside to bind in a separate pocket and been replaced by a protein His side chain, whereas the adenosyl group is not visible in the structure. The only other coenzyme B_{12}-containing enzyme of known structure, **methionine synthase** (Section 24-3a of this Supplement), has a similar B_{12}-binding fold in which the Co atom is likewise His-liganded, although this Co atom additionally has an CH_3 ligand opposite its His ligand. The Co atom in methylmalonyl-CoA mutase has no 6th ligand and hence, as confirmed by spectroscopic measurements, is in the Co(II) state. The His N—Co bond is extremely long (2.5 Å vs 1.9-2.0 Å in various other B_{12}-containing structures). It is proposed that this strained bond stabilizes the Co(II) state with respect to the Co(III) state, thus favoring the formation of the adenosyl radical and facilitating the homolytic cleavage through which the catalyzed reaction occurs (Fig. 23-20).

The desulfo-CoA binds to the α subunit's N-terminal domain with its pantetheine chain extended along a narrow tunnel down the center of the α/β barrel so as to put the methylmalonyl group of an intact substrate in close proximity to the top (unliganded) face of the cobalmin ring. This tunnel provides the only direct access to the active site cavity, thereby protecting the reactive free radical intermediates that are produced in the catalytic reaction from side reactions. The tunnel is lined by small hydrophilic residues (Ser and Thr), in contrast to the numerous other α/β barrel-containing enzymes in which the center of the barrel is generally occluded by large, often branched, hydrophobic side chains. Thus, methylmalonyl-CoA mutase's substrate binding mode is unique among α/β-barrel containing enzymes. Since the catalytically inactive β subunit has no obvious major function (the center of its α/β barrel is occluded), it seems likely that this subunit is an evolutionary relic.

Mancia, F., Keep, N.H., Nakagawa, A., Leadlay, P.F., McSweeney, S., Rasmussen, B., Bösecke, P., Diat, O., and Evans, P.R., How coenzyme B_{12} radicals are generated: the crystal structure of methylmalonyl-coenzyme A mutase at 2.0 Å resolution, *Structure* **4**, 339–350 (1996).

(b) A Ni—C Bond in an Enzymatic Reaction

The reaction catalyzed by the Ni-containing enzyme **carbon monoxide dehydrogenase** from the anaerobe *Clostridium thermoaceticum* combines a methyl group with CO to yield an acetyl group, which is then incorporated into acetyl-CoA. Resonance Raman spectroscopy reveals that the methyl group is provided by a Ni—CH_3 adduct, whose Ni—C bond represents only the second known instance of a biochemically significant metal—carbon bond; the first being the Co—C bond in coenzyme B_{12}.

Kumar, M., Qui, D., Spiro, T.G., and Ragsdale, S.W., A methylnickel intermediate in a bimetallic mechanism of acetyl-coenzyme A synthesis by anaerobic bacteria, *Science* **270,** 628–630 (1995).

(c) Additional References

Djordjevic, S., Pace, C.P., Stankovich, M.T., and Kim, J.-J., Three-dimensional structure of butyryl-CoA dehydrogenase from *Megasphaera elsdenii, Biochemistry* **34,** 2163–2171 (1995).

Leesong, M., Henderson, B.S., Gillig, J.R., Schwab, J.M., and Smith, J.L., Structure of a dehydratase–isomerase from the bacterial pathway for biosynthesis of unsaturated fatty acids: two catalytic activities in one active site, *Structure* **4,** 253–264 (1996). [The X-ray structure of *E. coli* β-hydroxydecanoyl thiol ester dehydratase.]

4. FATTY ACID BIOSYNTHESIS

Athappilly, F.K. and Hendrickson, W.A., Structure of the biotinyl domain of acetyl-coenzyme A carboxylase determined by MAD phasing, *Structure* **3,** 1407–1419 (1995). [The structure of *E. coli* biotin carboxyl-carrier protein, the first reported X-ray structure of a biotinyl domain (MAD is the acronym for *m*ultiwavelength *a*nomalous *d*iffraction, whose use in solving the X-ray structures of proteins was pioneered by Hendrickson).]

Rafferty, J.B., Simon, J.W., Baddock, C., Artymuik, P.J., Baker, P.J., Stuitje, A.R., Slabas, A.R., and Rice, D.W., Common themes in redox chemistry emerge from the X-ray structure of oilseed rape (*Brassica napus*) enoyl acyl carrier protein reductase, *Structure* **3,** 927–938 (1995).

Serre, L., Verbree, E.C., Dauter, Z., Stuitje, A.R., and Derwenda, Z.S., The *Eschericia coli* malonyl-CoA:acyl carrier protein transacylase at 1.4-Å resolution, *J. Biol. Chem.* **270,** 12961–12964 (1995).

Waldrop, G.L., Rayment, I., and Holden, H.M., Three-dimensional structure of the biotin carboxylase subunit of acetyl-CoA carboxylase, *Biochemistry* **33,** 10249–10256 (1994).

6. CHOLESTEROL BIOSYNTHEIS

Tarshis, L.C., Yan, M., Poulter, C.D., and Sacchetini, J.C., Crystal structure of a recombinant farnesyl diphosphate synthase at 2.6-Å, *Biochemistry* **33,** 10871–10877 (1994). [The X-ray structure of prenyl transferase.]

7.   ARACHIDONATE   METABOLISM: PROSTAGLANDINS, PROSTACYCLINS, THROMBOXANES, AND LEUKOTRIENES

DeWitt, D. and Smith, W.L., Yes, but do they still get headaches? *Cell* **83,** 345–348 (1995). [A review discussing the physiological consequences in mice of deleting either of their two cyclooxygenase activities.]

Loll, P., Picot, D., and Garavito, M., The structural basis of aspirin activity inferred from the crystal structure of inactivated prostaglandin H_2 synthase, *Nature Struct. Biol.* **2,** 637–643 (1995). [An X-ray structure showing that the acylation of Ser 530 of PGH_2 synthase by the aspirin analog 2-bromoacetoxy-benzoic acid abolishes the enzyme's cyclooxygenase activity, as was previously predicted, by steric blockage of the its active-site channel.]

Nelson, M.J. and Seitz, S.P., The structure and function of lipoxygenase, *Curr. Opin. Struct. Biol.* **4,** 878–884 (1994).

8. PHOSPHOLIPID AND GLYCOLIPID METABOLISM

Kent, C., Eukaryotic phospholipid biosynthesis, *Annu. Rev. Biochem.* **64,** 315–343 (1995).

Chapter 24

AMINO ACID METABOLISM

3. METABOLIC BREAKDOWN OF INDIVIDUAL AMINO ACIDS

(a) The Lipoamide Arm of the Glycine Cleavage System H Protein Is Not Freely Swinging

The glycine cleavage system (also known as glycine synthase and the **glycine decarboxylase complex**) functions to supply glycine's methylene group to THF to form N^5,N^{10}-methylene-THF, with the remainder of the glycine released as CO_2 and NH_4^+ (Fig. 24-11). It is a 4-component multienzyme complex that resembles the pyruvate dehydrogenase multienzyme complex (Section 19-2A). The H protein of the glycine cleavage system contains a lipoamide prosthetic group that transfers glycine's aminomethyl group from a PLP-dependent glycine decarboxylase (P protein) to an N^5,N^{10}-methylene-THF synthesizing enzyme (T protein) in much the same way that the lipoamide group on E_2 of the pyruvate dehydrogenase complex transfers pyruvate's acetyl group from pyruvate dehydrogenase (E_1) to CoA (Fig. 19-6). In both complexes, the resulting dihydrolipoyl group is reoxidized to lipoate by an NAD^+-dependent flavoenzyme (L-protein in the glycine cleavage system and E_3 in the pyruvate dehydrogenase complex).

The X-ray structures were determined of the 131-residue pea leaf mitochondrial H protein in its oxidized (lipoamide) form and in two different types of crystals of its reduced form in which the aminomethyl intermediate is covalently linked to the dihydrolipoamide group. H protein is largely comprised of a sandwich of a 3-stranded and a 6-stranded antiparallel β sheet that is structurally similar to the lipoyl domain of the pyruvate dehydrogenase complex (Section 19-2b of this Supplement). In both structures, the lipoic acid is linked via a Schiff base to a Lys residue (the lipollysyl arm; Fig. 19-8) at the tip of a tight turn between two β strands.

In the pyruvate dehydrogenase complex, the 14-Å long lipollysyl arm, together with polypeptide segment of E_2, appear to be freely swinging so as to convey the lipoamide group between the active sites of E_1, E_2, and E_3. It had therefore been assumed that H protein's lipollysyl arm acts in a similar manner. Indeed, the X-ray structure of the oxidized H protein shows its lipollysyl arm bound to the surface of the enzyme in what appears to be a flexible conformation. However, in both crystal structures of the reduced form, the lipollysyl arm, together with its covalently linked aminomethyl group, make a U-turn to tuck into a hydrophobic cleft on the surface of the H protein, in which the aminomethyl group forms 3 hydrogen bonds. This cleft, it is postulated, protects the reactive aminomethyl group from hydrolysis to form NH_4^+ and formaldehyde. Thus, the lipollysyl– aminomethyl arm is not free to move about in the aqueous solvent.

Pares, S., Cohen-Addad, C., Sieker, L., Neuburger, M., and Douce, R., X-ray structure determination at 2.6-Å resolution of a lipoate containing protein: The H-protein of the glycine decarboxylase complex from pea leaves, *Proc. Natl. Acad. Sci.* **91,** 4850–4853 (1994); The lipoamide arm of the glycine decarboxylase complex is not freely swinging, *Nature Struct. Biol.* **2,** 63–68 (1995); *and* Refined structures at 2 and 2.2 Å resolution of two forms of the H-protein, a lipoamide-containing protein of the glycine decarboxylase complex, *Acta Cryst.* **D51,** 1041–1051 (1995).

(b) Structure of Methionine Synthase, a Coenzyme B_{12}-Dependent Methyltransferase

The enzyme catalyzing the methylation of homocysteine by N^5-methyl-THF, homocysteine methyltransferase (alternatively, **methionine synthase),** is the only coenzyme B_{12}-associated enzyme in mammals besides methylmalonyl-CoA mutase (Section 23-2E). However, the coenzyme B_{12}'s Co ion in homocysteine methyltransferase is axially liganded by a methyl group to form methylcobalamin rather than by a 5'-deoxyadenosyl group present in methylmalonyl-CoA mutase.

The X-ray structure of the 246-residue methylcobalamin-binding segment of the 1227 monomeric *E. coli* homocysteine methyltransferase, the first known structure of a coenzyme B_{12}-binding protein, reveals it to consist of two domains: an N-terminal helical domain and a C-terminal α/β domain that is a variant of the Rossmann fold (Section 7-3B) and whose core consists of a 5-stranded parallel β sheet. The corrin ring is sandwiched between these domains. Contrary to previous expectation, the coenzyme's 5,6-dimethylbenzamidazole (DMB) moiety (Fig. 23-19) is not liganded to the Co ion. Rather, the Co ion is liganded by an enzyme His side chain and the DMB is anchored between two β strands at some distance from the corrin ring. Sequence homologies suggest that the forgoing Rossmann fold variant may be a common binding motif in B_{12}-associated enzymes. Indeed, the B_{12}-binding site in the subsequently determined X-ray structure of methylmalonyl-CoA mutase (Section 23-2a of this Supplement) has a similar fold and its DMB group has also swung aside to be replaced by a protein His ligand to the Co ion.

Drennan, C.L.,Huang, S., Drummond, J.T., Matthews, R.G., and Ludwig, M.L., How a protein binds B_{12}: A 3.0 Å X-ray structure of B_{12}-binding domains of methionine synthase, *Science* **266,** 1669–1674 (1994).

Drennan, C.L., Matthews, R.G., and Ludwig, M.L., Cobalamin-dependent methionine synthase:

the structure of a methylcobalamin-binding fragment and implications for other B_{12}-dependent enzymes, *Curr. Opin. Struct. Biol.* **4,** 919–929 (1994).

(c) Additional References

Baker, P.J., Turnbull, A.P., Sedelnikova, S.E., Stillman, T.J., and Rice, D.W., A role for quaternary structure in the substrate specificity of leucine dehydrogenase, *Structure* **3,** 693–705 (1995). [Reports the X-ray structure of the enzyme catalyzing Reaction 1C in the leucine degradation pathway (Fig. 24-15). Leucine dehydrogenase is an NAD^+-linked homo-octameric enzyme that has a similar catalytic mechanism to that of glutamate dehydrogenase, a 20% identical sequence, and a homologous tertiary structure.]

4. AMINO ACIDS AS BIOSYNTHETIC PRECURSORS

(a) Malarial Hemozoin Polymerization Is Not Enzyme-Catalyzed

Malarial parasites digest up to 80% of the hemoglobin in their host erythrocytes. The potentially toxic heme groups released by this process are sequestered into insoluble deposits of hemozoin, which consists of polymerized oxidized hemes linked by iron-carboxylate bonds between their Fe(III) atoms and the propionate side chains of adjacent molecules.

The so-called ''heme polymerase'' that, as reported in the textbook, had been implicated in catalyzing this process and was thought to be inhibited by the antimalarial drug chloroquine has now been shown to be nonexistent. Rather, heme polymerization has been demonstrated to be a chemical process that is catalyzed by hemozoin itself rather than by any protein. This autocatalytic process is, nevertheless, still inhibited by chloroquine and hence remains a target for new antimalarial drugs. The way in which hemozoin polymerization is initiated remains unknown.

Dorn, A., Stoffel, R., Matile, H., Bubendorf, A., and Ridley, R.G., Malarial haemozoin/β-haematin supports haem polymerization in the absence of protein, *Nature* **374,** 269–271 (1995).

(b) Additional References

Brownlee, P.D., Lambert, R., Louie, G.V., Jordan, P.M., Blundell, T.L., Warren, M.J., Cooper, J.B., and Wood, S.P., The three-dimensional structures of mutants of porphobilinogen deaminase: Toward and understanding of the structural basis of acute intermittent porphyria, *Protein Science* **3,** 1644–1650 (1994).

Jordan, P.M., Highlights in haem biosynthesis, *Curr. Opin. Struct. Biol.* **4,** 902–911 (1994).

May, B.K., Dogra, S.C., Sadlon, T.J., Bhasker, C.R., Cox, T.C., and Bottomley, S.S., Molecular regulation of heme biosynthesis in higher vertebrates, *Prog. Nucl. Acid Res. Mol. Biol.* **51,** 1–51 (1995).

5. AMINO ACID BIOSYNTHESIS

Cheah, E., Carr, P.D., Suffolk, P.M., Vasudevan, S.G., Dixon, N.E., and Ollis, D.L., Structure of the *Eschericia coli* signal transducing protein P_{II}. *Structure* **2**, 981–980 (1994). [P_{II}, the regulatory enzyme that adenylylates and deadenylylates glutamine synthase in *E. coli,* is a trimer of identical 112-residue subunits. Its Tyr 51, which is uridylylated by uridylyltransferase, is located at the end of a loop projecting from one face of the trimer and is therefore in a solvent-exposed position that is ideally suited for signal transduction.]

Hurley, J.H. and Dean, A.M., Structure of 3-isopropylmalate dehydrogenase in complex with NAD^+: ligand-induced loop closing and mechanism for cofactor specificity, *Structure* **2**, 1007–1016 (1994). [The enzyme that catalyzes Reaction 8 in the leucine synthesis pathway (Fig. 24-48). It is structurally and mechanistically related to the citric acid cycle enzyme isocitrate dehydrogenase.]

Mirwaldt, C., Korndörfer, I., and Huber, R., The crystal structure of dihydrodipicolinate synthase from *Eschericia coli* at 2.5 Å resolution, *J. Mol. Biol.* **246**, 227–239 (1995). [The enzyme that catalyzes Reaction 10 in the lysine synthesis pathway (Fig. 24-27; in which dihydrodipicolinate and dihydrodipicolinate synthase are erroneously named dihydropicolinate and dihydropicolinate synthase).]

Scapin, G., Blanchard, J.S., and Sacchettini, J.C., Three-dimensional structure of *Eschericia coli* dihydrodipicolinate reductase, *Biochemistry* **34**, 3502–3512 (1995). [The enzyme that catalyzes Reaction 11 in the lysine synthesis pathway (Fig. 24-27).]

Chapter 25

ENERGY METABOLISM: INTEGRATION AND ORGAN SPECIALIZATION

2. ORGAN SPECIALIZATION

(a) Leptin and its Receptor Regulate Appetite and Body Weight

When a normal animal overeats, the resulting additional adipose (fat) tissue somehow signals the brain to induce the animal to eat less and to expend more energy. Conversely, the loss of fat stimulates increased eating until the lost fat is replaced. Evidently, animals have a ''lipostat'' that can keep the amount of body fat constant to within 1% over many years. At least a portion of the lipostat resides in the hypothalamus (a part of the brain that hormonally controls numerous physiological functions; Section 34-4A) since damage to it can induce gross obesity.

Mutations at five genetic loci are known to cause obesity in mice. Of these, the two that have received the most attention are known as *obese* (*ob*) and *diabetes* (*db*). Homozygotes for defects in either of these recessive genes, *ob/ob* and *db/db,* are grossly obese and have nearly identical phenotypes. Indeed, the only way that these phenotypes had been distinguished was by surgically linking the circulation of a mutant mouse to that of a normal mouse, a phenomenon named **parabiosis.** *ob/ob* mice so-linked exhibited normalization of body weight and reduced food intake, whereas *db/db* mice did not do so. This suggests that the *ob/ob* mice are deficient in a circulating factor that regulates appetite and metabolism, whereas *db/db* mice are defective in the receptor for this circulating factor.

The mouse *ob* gene has been cloned and shown to encode a 167-residue protein, named **leptin** (Greek: *leptos,* thin), that has no apparent homology with proteins of known sequence. Leptin is expressed only by adipocytes (fat cells), which in doing so appear to inform the brain of how much fat the body carries. Thus, injecting leptin into *ob/ob* mice causes them to eat less and to lose weight. In fact, leptin-treated *ob/ob* mice on a restricted diet lost 50% more weight than untreated *ob/ob* mice on the same diet, which suggests that reduced food intake alone is insufficient to account for leptin-induced weight loss. However, leptin injection has no effect on *db/db* mice.

The leptin receptor gene was identified by making a cDNA library from mouse brain tissue that specifically bound leptin and then identifying a receptor-expressing clone by its ability to bind leptin (gene cloning techniques are discussed in Section 28-8). This gene, which has been shown to be the *db* gene, encodes a protein, named **OB-R**, that appears to have a single transmembrane segment and an extracellular leptin-binding domain that resembles the receptors for certain cytokines (proteins that regulate the differentiation, proliferation, and activities of various blood cells; Section 34-4B).

OB-R protein has at least six alternatively spliced forms that appear to be expressed in a tissue-specific manner (alternative gene splicing is discussed in Section 33-3C). In normal mice, the hypothalamus expresses, at high levels, a splice variant of OB-R that has a 302-residue cytoplasmic segment. However, in *db/db* mice, this segment has an abnormal splice site that truncates it to only 34 residues, which almost certainly renders this OB-R variant unable to intracellularly transmit the signal that it has extracellularly bound leptin. Thus, it appears that leptin's weight-controlling effects are mediated by signal transduction resulting from its binding to the OB-R protein in the hypothalamus (signal transduction is discussed in Section 34-4B).

Human leptin is 84% identical in sequence to that of mice. The use of a radioimmunoassay (Section 34-4A) to measure the serum levels of leptin in normal-weight and obese humans revealed that their serum leptin concentrations increase with their percentage of body fat as does the *ob* mRNA content of their adipocytes. Moreover, after obese individuals had lost weight, their serum leptin concentrations and adipocyte *ob* mRNA content declined. This suggests that most obese persons produce sufficient amounts of leptin but are insensitive to its presence, perhaps in a manner similar to that of a *db/db* mouse.

Zhang, Y., Proenca, R., Maffei, M., Barone, M., Leopold, L., and Friedman, J.M., Positional cloning of the mouse *obese* gene and its human homologue, *Nature* **372,** 425–432 (1994).

Tartaglia, L.A., et al., Identification and expression cloning of a leptin receptor, OB-R, *Cell* **83,** 1263–1271 (1995).

Lee, G.-H., Proenca, R., Montez, J.M., Carroll, K.M., Darvishzadah, J.G., Lee, J.I., and Friedman, J.M., Abnormal splicing of the leptin receptor in *diabetic* mice, *Nature* **379,** 632–635 (1996).

Chen, H., et al., Evidence that the diabetes gene encodes the leptin receptor: Identification of a mutation in the leptin receptor gene in *db/db* mice, *Cell* **84,** 491–495 (1996); *and* Chua, S.C., Jr., Chung, W.K., Wu-Peng, S., Zhang, Y., Liu, S.-M., Tartaglia, L., and Leibel, R.L., Phenotypes of mouse *diabetes* and rat *fatty* due to mutations in the OB (leptin) receptor, *Science* **271,** 994–996 (1996).

Considine, R.V., et al., Serum immunoreactive-leptin concentrations in normal-weight and obese humans, *New Engl. J. Med.* **334,** 292–295 (1996).

3. METABOLIC ADAPTATION

Newgard, C.B. and McGarry, J.D., Metabolic factors in pancreatic β-cell signal transduction, *Annu. Rev. Biochem.* **64,** 689–719 (1995).[Reviews the biochemical mechanisms that mediate glucose-stimulated insulin secretion by pancreatic β cells.]

Chapter 26

NUCLEOTIDE METABOLISM

2. SYNTHESIS OF PURINE RIBONUCLEOTIDES

(a) X-ray Structure of Amidophosphoribosyl Transferase in Complex with AMP

Amidophosphoribosyl transferase (also known as **glutamine PRPP amidotransferase**) catalyzes Reaction 2 of the pathway for the *de novo* synthesis of purine nucleotides (Fig. 26-3), the first committed step in this 11-step pathway. The enzyme is the only member of this pathway that is regulated by purine nucleotides (AMP, ADP, GMP, and GDP). The amidophosphoribosyl transferases from many, but not all, organisms contain a [4Fe–4S] cluster that has a regulatory rather than a catalytic function: Reaction with O_2 causes cluster decomposition, which in turn, irreversibly denatures the enzyme. It has therefore been proposed that the [4Fe–4S] cluster acts as an O_2 sensor that detects nutrient limitation and thereby stimulates proteolytic degradation (recall that the citric acid cycle enzyme aconitase has a [4Fe–4S] cluster that, although it has a catalytic function, does not participate in the redox reactions normally associated with iron–sulfur clusters; Section 19-3B).

The X-ray structure of amidophosphoribosyl transferase from *Bacillus subtilis* in complex with its allosteric inhibitor AMP reveals that this [4Fe–4S] cluster-containing tetramer of identical 465-residue subunits binds two AMPs per subunit. Each subunit binds a [4Fe–4S] cluster in a way that shields the cluster from the bulk solvent except for one of its S^{2-} ions, which is solvent-accessible through a 7-Å wide × 7-Å deep channel. AMP strongly protects the *B. subtilis* enzyme from O_2-dependent degradation. In fact, one of the two AMPs associated with each subunit, that which occupies the so-called R (for regulatory) site, seals over the channel through which the [4Fe–4S] cluster would otherwise be exposed. The ribose-5'-phosphate moiety of the second AMP, which occupies the enzyme's so-called C (for catalytic) site, probably mimics its position in the enzyme-bound PRPP substrate. An AMP bound in the C site would clearly inhibit the enzyme by direct steric blockage. This also seems to be the case for an AMP occupying the R site because its 5'-phosphate group would overlap the α-pyrophosphoryl group of a PRPP bound to the C site in the same way as is AMP. This is an unexpected observation because, according to biochemical criteria, amidophosphoribosyl transferase is a classic allosteric enzyme exhibiting no competitive inhibition by AMP. However, since the PRPP-binding site is too far away from the enzyme's putative glutamine-binding site for effective catalysis, it appears that a large conformational change is required to bring together the protein's two substrate-binding sites to form the catalytically active enzyme.

Smith, J.L., Zaluzec, E.J., Wery, J.P., Niu, L., Switzer, R.L., Zalkin, H., and Satow, Y., Structure of the allosteric regulatory enzyme of purine biosynthesis, *Science* **264,** 1427–1433 (1994).

(b) Additional References

Klein, C., Chen, P., Arevalo, J.H., Stura, E.A., Marolewski, A., Warren, M.S., Benkovic, S.J., and Wilson, I.A., Towards structure-based drug design: Crystal structure of a multisubstrate adduct complex of glycinamide ribonucleotide transformylase at 1.96 Å resolution, *J. Mol. Biol.* **249,** 153–179 (1995). [The X-ray structure of *E. coli* GAR transformylase, the enzyme catalyzing Reaction 4 in the pathway synthesizing IMP (Fig. 26-3) in complex with a bisubstrate analog. The structure of GAR transformylase in complex with GAR and the inhibitor 5-dTHF is shown in Fig. 26-4.]

Silva, M.M., Poland, B.W., Hoffman, C.R., Fromm, H.J., and Honzatko, R.B., Refined crystal structures of unligated adenylosuccinate synthetase from *Eschericia coli, J. Mol. Biol.* **254,** 431–446 (1995); *and* Wang, W., Poland, B.W., Honzatko, R.B., and Fromm, H.J., Identification of arginine residues in the putative L-aspartate binding site of *Eschericia coli* adenylolsuccinate synthetase, *J. Biol. Chem.* **270,** 13160–13163 (1995). [The structure of the first of the two enzymes that convert IMP to AMP (Fig. 26-6).]

Smith, J.L., Enzymes of nucleotide synthesis, *Curr. Opin. Struct. Biol.* **5,** 752–757 (1995).

3. SYNTHESIS OF PYRIMIDINE RIBONUCLEOTIDES

Scapin, G., Ozturk, D.H., Grubmeyer, C., and Sacchettini, J.C., The crystal structure of orotate phosphoribosyltransferase complexed with orotate and α-D-phosphoribosyl-1-pyrophosphate, *Biochemistry* **34,** 10744–10754 (1995). [The enzyme catalyzing Reaction 5 of the pathway synthesizing UMP (Fig. 26-9).]

4. FORMATION OF DEOXYRIBONUCLEOTIDES

(a) Inhibitory C-Terminal Peptides from the R2 Subunit of Ribonucleotide Reductase As Potential Chemotherapeutic Agents

Ribonucleotide reductase catalyzes an essential reaction in the *de novo* synthesis of DNA, the reduction of the C2'-hydroxyl group of NDPs to yield dNDPs. Hence, ribonucleotide reductases are potential chemotherapeutic targets for the treatment of cancers as well as viral and parasitic diseases. Class I ribonucleotide reductases consist of two homodimeric subunits, R1 (a dimer of α subunits) and R2 (a dimer of β subunits). Each subunit of R2, whose X-ray structure is shown in Fig. 26-12*b,* contains an Fe(III) atom and a stable tyrosyl radical. Each subunit of R1, whose X-ray structure is shown in Fig. 26-12d, binds the substrate NDP and, in addition, contains two allosteric nucleoside triphosphate binding sites. Catalytic activity requires the transfer of the radical character from the tyrosyl radical in R2 to the substrate bound to R1 (Fig. 26-13), and consequently, that R1 and R2 be properly assembled.

R1 specifically binds to R2 via the latter's C-terminal peptide. Indeed, *E. coli* R1 could only be satisfactorily crystallized in complex with the 20-residue C-terminal peptide of *E. coli* R2. Since the R2 C-terminal peptides from different species vary in their sequences, species-specific peptides or nonpeptide compounds that have similar conformations and binding properties (compounds known as **peptidomimetics**) would inhibit the assembly of their corresponding ribonucleotide reductase and hence are potential chemotherapeutic agents.

The C-terminal heptapeptide of mouse R2, Phe-Thr-Leu-Asp-Ala-Asp-Phe (FTLDADF), is the minimal peptide that can inhibit mouse ribonucleotide reductase. Thus, the conformations of the N-acetylated form of this peptide (N-AcFTLDADF) and two of its variants (N-AcYTLDADF and N-AcFTLDADL), when bound to mouse R1, were determined in solution using NMR techniques. The most reliable structure, that of N-AcYTLDADF, had as its most prominent structural feature a reverse turn formed by its TLDA segment. This closely matches the X-ray structure-derived conformation of the LNSF segment of the C-terminal octapeptide, DDLSNFDL, of *E. coli* R2 when it is bound to *E. coli* R1. However, outside these segments, the two peptides diverged in conformation. This suggests that species-specific sequences can be exploited in the design of inhibitors directed against the ribonucleotide reductases from specific organisms.

Fisher, A., Laub, P.B., and Cooperman, B.S., NMR structure of an inhibitory R2 C-terminal peptide bound to mouse ribonucleotide reductase R1 subunit, *Nature Struct. Biol.* **2,** 951–955 (1995).

(b) Additional References

Carreras, C.W. and Santi, D.V., The catalytic mechanism and structure of thymidylate synthase, *Annu. Rev. Biochem.* **64,** 721–762 (1995).

Licht, S., Gerfen, G.J., and Stubbe, J., Thiyl radicals in ribonucleotide reductases, *Science* **271,** 477–481 (1996). [EPR measurements indicate that the Class II ribonucleotide reductase from *Lactobacillus leichmannii* (which has a 5'-deoxyadenosylcobalmin prosthetic group) has a thiyl radical intermediate that initiates the reduction process as has previously been established for Class I ribonucleotide reductases (whose active sites contain Fe(III)-stabilized tyrosyl radicals).]

Martin, J.L., Thioredoxin – a fold for all reasons, *Structure* **3,** 245–250 (1995).

5. NUCLEOTIDE DEGRADATION

(a) X-Ray Structure of a Xanthine Oxidase-Related Aldehyde Oxido-Reductase

The aldehyde oxido-reductase (**Mop**) from the sulfate-reducing anaerobic bacterium *Desulfovibrio gigas* oxidizes aldehydes to carboxylic acids with little specificity for the nature of the side chain. Mop is a dimer of 907-residue subunits that are 26% identical (52% similar) in sequence to xanthine oxidase from *D. melanogaster,* which is also a homodimer. As does xanthine oxidase, each Mop subunit binds two [2Fe–2S] clusters and a molybdenum-containing cofacter named **Mo-co.** However, Mop lacks xanthine oxidase's FAD cofactor (xanthine oxidase subunits consist of ~1300 residues).

The X-ray structure of Mop reveals that each of its subunits folds into four domains, the first two of which each bind one of the [2Fe–2S] clusters with the remaining domains binding the Mo-co. Mo-co consists of a Mo ion complexed to a pterin derivative named **molybdopterin** that is linked to the nucleoside diphosphate CDP. Referring to the diagram of pterin in Fig. 24-20, molybdopterin–CDP consists of pterin with the four-carbon R group $-C(SH)=C(SH)-CH(O-)-CH_2-O-CDP$. The S atoms of the cis-enedithiol group at C7' and C8' coordinate the Mo ion. The O atom substituent at C9' is also covalently bonded to C7 of the pterin ring, thereby forming a pyran ring (and hence Mo-co is a tricyclic system). C10' forms a phosphoester bond to the β phosphate of CDP. The Mo ion is additionally coordinated by three poorly identified ligands that appear to be O atoms. The Mo-co is deeply buried in the protein but is accessible to the outside via 15-Å long funnel-shaped tunnel that leads to the Mo ion.

Romão, M.J., Archer, M., Moura, I., Moura, J.J.G., LeGall, J., Engh, R., Schneider, M., Hof, P., and Huber, R., Crystal structure of the xanthine-oxidase–related aldehyde oxidoreductase from D. gigas, *Science* **270,** 1170-1176 (1995).

(b) Additional References

Eads, J.C., Scapin, G., Xu, Y., Grubmeyer, C., and Sacchettini, J.C., The crystal structure of human hypoxanthine-guanine phosphoribosyltransferase with bound GMP, *Cell* **78,** 325–334 (1994).

Chapter 28

NUCLEIC ACID STRUCTURES AND MANIPULATION

2. DOUBLE HELICAL STRUCTURES

(a) Structures of RNA and DNA Double Helices with the Same Sequence

The structures of few double helical RNA segments, besides those forming the stems of tRNAs (Sections 30-2B and C) and 5S ribosomal RNA (Fig. 30-29), have been reported. Recently, the X-ray structure of r(CCCCGGGG) has been determined in two crystal lattices. The self-complementary RNAs in the two crystal forms are closely similar A-type helices that have ~10.6 bp per turn, a pitch (rise per turn) of ~26.5 Å, and only small deviations from regular geometry. The comparison of these helices with that in the previously determined structure of a A-DNA segment that has the identical base sequence, d(CCCCGGGG), reveals that these helices assume distinctly different conformations. The DNA helix has a wider major groove than does the RNA (~9.2 Å vs ~3.9 Å), a smaller inclination of its base pairs to the helix axis (~8.0° vs ~16°), and a pitch of ~33.9 Å which makes it a longer, slimmer helix than that of the RNA.

In the several reported cases in which the same DNA segment crystallizes in two different crystal forms or occupies two nonequivalent sites in the same crystal lattice, the DNAs exhibit significantly different conformations, which suggests that they are flexible. In contrast, the RNA double helices in the two r(CCCCGGGG) structures, as well as the corresponding helical stems of chemically identical tRNAs that crystallize in different lattices, show little conformational variability. This suggests that RNA double helices are more rigid than DNA helices and hence that proteins which bind to RNA double helices will not distort the conformations of their target RNAs to the same degree that many DNA-binding proteins distort their target DNAs [e.g., in the DNA complexes of *Eco*RV restriction endonuclease (Fig. 28-49b) and TATA box-binding protein (Fig. 33-48a)].

Portmann, S., Usman, N., and Egli, M., The crystal structure of r(CCCCGGGG) in two distinct lattices, *Biochemistry* **34,** 7569–7575 (1995).

(b) X-Ray Structures of RNA Double Helices Containing Non-Watson-Crick Base Pairs

In the RNA dodecamer GGAC<u>UUUG</u>GUCC, the central four bp (*underlined*) are not self-complementary in the Watson-Crick sense. Nevertheless, the X-ray structure of this RNA segment reveals that it forms a double helix whose entire sugar-phosphate backbone closely follows that of the cannonical (standard) A-RNA structure. The central segment contains two U·G base pairs and two U·U base pairs. The U·G base pairs have the standard "Wobble" geometry (Fig. 30-24). In one of the U·U base pairs, the N3—H and O2 groups of one U hydrogen bond to the O4 and the N3—H groups of the other U, respectively. The second U·U base pair has a similar geometry but is propeller-twisted by ~45° such that its N3—H···O2 hydrogen bond does not form. However, this propeller twisting brings together the two U-O4 atoms in adjacent U·U base pairs that are not hydrogen bonded to an N3—H group such that they are bridged by a water molecule in the major groove, thereby compensating for the loss of the N3—H···O2 hydrogen bond. A similar U·U base pair may form between U(−1) and U(120) of the *E. coli* 5S RNA (bottom of Fig. 30-29b) but the NMR structure of its Helix I is of insufficient definition in this region to unequivocally establish this to be the case.

In a similar study, the X-ray structure of r(CGC<u>G</u>AAUU<u>A</u>GCG) reveals that this mostly self-complementary RNA forms a typical A-type duplex that contains two G·A base pairs in which the N1—H and O6 groups of each G respectively form hydrogen bonds with the N1 and N6-amino groups of its paired A, much like that diagrammed in the lower center of Fig. 30-15. Surprisingly, the G·A base pairs, which are wider than Watson-Crick base pairs, appear to be incorporated into the A-RNA helix with only a slight bulging of its sugar-phosphate backbone in the immediate vicinity of the G·A pairs and with little alteration in its conformational angles.

Baeyens, K.J., De Bondt, H.L., and Holbrook, S.R., Structure of an RNA double helix including uracil-uracil base pairs in an internal loop, *Nature Struct. Biol.* **2,** 56–62 (1995).

Leonard, G.A., McAuley-Hecht, K.E., Ebel, S., Lough, D.M., Brown, D.M., and Hunter, W.N., Crystal and molecular structure of r(CGCGAAUUAGCG): an RNA duplex containing two G(*anti*)·A(*anti*) base pairs, *Structure* **2,** 483–494 (1994).

(c) Triple-Helical Base Pairing

Triple helices, in which the bases of a third strand are Hoogsteen-paired to the bases of a Watson-Crick double helix, have been observed in fiber diffraction studies such as those of poly(U)·poly(A)*poly(U) [where the asterisk (*) indicates the Hoogsteen pairing interactions]. Such interactions may occur during homologous recombination (Section 31-6A) and may provide a basis for the artificial control of gene activity.

In the X-ray structure of the mostly self-complementary nonamer d(GCGAATTCG), the overhanging G on each of two B-DNA double helices pairs with the terminal Watson-Crick C·G base pair on the other helix to form two (C·G)*G base triplets. One of these G·G interactions has a Hoogsteen-like geometry with N1—H···O6 and N2—H···N7 hydrogen bonds (as is seen in the G-quartet diagrammed in Fig. 31-35a), whereas the other has the reversed geometry in that it is linked by N1—H···N7 and N2—H···O6 hydrogen bonds (as seen in the structure of yeast tRNA^Phe; lower left of Fig. 30-15). In the former C·G*G interaction, O6 of the non-Watson-Crick paired G is within hydrogen bonding distance (2.93 Å) of the C's N4 amino group although the ~110° angle of this N—H···O interaction deviates far from the linear geometry of an ideal hydrogen bond.

Van Meervelt, L., Vlieghe, D., Dautant, A., Gallois, B., Précigoux, G., and Kennard, O., High resolution structure of a DNA helix forming (C·G)*G base triplets, *Nature* **374,** 742–744 (1995).

(d) Additional References

Betts, L., Josey, J.A., Veal, J.M., and Jordan, S.R., A nucleic acid triple helix formed by a peptide nucleic acid–DNA complex, *Science* **270,** 1838–1841 (1995). [The X-ray structure of a nucleic acid triple helix formed by a 9-nt polypurine DNA strand complexed with a polypyrimidine **peptide nucleic acid (PNA;** a synthetic polypeptide consisting of δ-amino acid residues whose uncharged side chains bear nucleic acid bases) that is folded into a hairpin.]

Frank-Kamenetskii, M.D. and Mirkin, S.M., Triplex DNA structures, *Annu. Rev. Biochem.* **64,** 65–95 (1995).

Gold, L., Polisky, B., Uhlenbeck, O., and Yarus, M., Diversity of oligonucleotide functions, *Annu. Rev. Biochem.* **64,** 763–797 (1995).[Describes a technology known as SELEX for the identification of high affinity oligonucleotide ligands. The resulting oligonucleotides, termed aptamers, have a variety of chemical and biomedical applications.]

Ji, J., Hogan, M.E., and Gao, X., Solution structure of an antiparallel purine motif triplex containg a *T*·CG pyrimidine base triple, *Structure* **4,** 425–435 (1996).

Neidle, S., *DNA Structure and Recognition,* IRL Press (1994).

Nicholson, A.W., Structure, reactivity, and biology of double-stranded RNA, *Prog. Nucleic Acid Res. Mol. Biol.* **52,** 1–65 (1996).

Pyle, A.M. and Green, J.B., RNA folding, *Curr. Opin. Struct. Biol.* **5,** 303–310 (1995).

Schindelin, H., Zhang, M., Bald, R., Fürste, J.-P., Erdmann, V.A., and Heinemann, U., Crystal structure of an RNA dodecamer containing the *Eschericia coli* Shine-Dalgarno sequence, *J. Mol. Biol.* **249,** 595–603 (1995).

Sinden, R.R., *DNA Structure and Function,* Academic Press (1994).

Varani, G., Exceptionally stable nucleic acid hairpins, *Annu. Rev. Biophys. Biomol. Struct.* **24,** 379–404 (1995).

Voet, D. and Voet, J.G., *KINEMAGES to Accompany Biochemistry, 2/E,* Wiley (1996). [Exercise 18 contains kinemages showing and comparing the structures of A-, B-, and Z-DNAs.]

3. FORCES STABILIZING NUCLEIC ACID STRUCTURES

Plum, G.E., Pilch, D.S., Singleton, S.F., and Breslauer, K.J., Nucleic acid hybridization: Triplex stability and energetics, *Annu. Rev. Biophys. Biomol. Struct.* **24,** 319–350 (1995).

Sharp, K.A. and Honig, B., Salt effects on nucleic acids, *Curr. Opin. Struct. Biol.* **5,** 323–328 (1995).

Strauss, J.K. and Maher, L.J., III, DNA bending by asymmetric phosphate neutralization, *Science* **266,** 1829–1834 (1994). [If the anionic charges on one face of a DNA molecule are neutralized, those on the opposite face, through charge-charge repulsions, should cause the DNA to bend toward the direction of the neutral face. This has been shown to be the case by replacing selected phosphate groups in DNA by neutral methylphosphonate groups. This suggests that proteins that bend their bound DNAs around them (e.g., CAP; Fig. 29-22a) may do so by placing positively charged side chains in juxtaposition with the phosphate groups along one face of the DNA.]

5. SUPERCOILED DNA

(a) X-Ray Structure of a 26-kD Fragment of Yeast DNA Topoisomerase I

Type I DNA topoisomerases catalyze the topological transformation of DNAs through the transient cleavage of one DNA strand, the passage of a second strand through the break, and the resealing of the break (Fig. 28-41), thereby incrementing the DNA's linking number by 1. There are two subfamilies of these proteins. The first, which includes *E. coli* topoisomerase I (the structure of whose N-terminal DNA cleavage domain is shown in Fig. 28-42), preferentially relaxes negatively supercoiled DNA and forms a covalent 5'-phosphotyrosine intermediate that preserves the free energy of the cleaved phosphodiester bond for subsequent rejoining. The other, which includes eukaryotic type I topoisomerases such as yeast topoisomerase I, relaxes both negatively and positively supercoiled DNAs and forms transient 3'-phosphotyrosine intermediates.

Yeast (*S. cerevisiae*) topoisomerase I does not require its N-terminal 135-residue segment to relax supercoiled DNA (although it is necessary for the *in vivo* complementation of *E. coli* topoisomerase I). The X-ray structure of a segment of yeast topoisomerase I comprising its residues 135 to 363 reveals that it consists of 2 domains that form a V-shaped concave platform that exhibits no structural similarity to *E. coli* topoisomerase I. The concave platform has a positively charged surface and a ~30-Å diameter of curvature, which is similar to those of many DNA-binding proteins [e.g., *E. coli* DNA topoisomerase I and TATA box-binding protein (Fig. 33-48)]. Although the 26-kD fragment does not exhibit significant DNA-binding affinity, the presence of DNA nevertheless protects the several Lys residues that occupy its positively charged concave platform from derivitization by citraconic anhydride (Section 6-1E). This has prompted the formulation of a model in which B-DNA binds to this surface.

Camptothecin is an antitumor alkaloid that stabilizes the covalent intermediate between DNA and eukaryotic topoisomerase I. Two residues, whose mutations reduce the sensitivity of mammalian topoisomerse I to camptothecin, map to a loop on yeast topoisomerse I that forms the base of the V-shaped platform. This suggests that this loop is near to or part of the enzyme's active site for DNA cleavage and rejoining.

Lue, N., Sharma, A., Mondragón, A., and Wang, J.C., A 26 kDa yeast DNA topoisomerase I fragment: crystallographic structure and mechanistic implications, *Structure* **3**, 1315–1322 (1995).

(b) X-Ray Structure of a Large Fragment of Yeast Topoisomerase II

Type II DNA topoisomerases (un)wind supercoiled DNAs through a transient double strand cleavage, the passage of the duplex through the break, and the resealing of the break (Figs. 28-43 and 28-44), thereby incrementing the DNA's linking number by 2. Yeast *(S. cerevisiae)* topoisomerase II is a type II topoisomerase whose sequence is homologous to that of the B subunit of *E. coli* gyrase followed by its A subunit. The 92-kD segment encompassing residues 410 to 1202 of this 1429-residue protein can cleave duplex DNA but cannot transport it through the break because it lacks the enzyme's ATPase domain (residues 1-409). However, the C-terminal segment (residues 1203-1429) appears dispensable .

The X-ray structure of the 92-kD segment reveals that has the shape of a flattened crescent and consists of two subfragments that are designated B' (residues 420-633), and A' (residues 682-1178), due to their homology to the B and A subunits of *E. coli* gyrase. The 48-residue segment between these two subfragments is disordered.

Two crescent-shaped monomers associate to form a heart-shaped dimer with its two B' subfragments associating at the top of the heart and its two A' subfragments coming together at its base (point). The dimer encloses a large triangular central hole (55 Å wide at its base and 60 Å in height), thereby forming a structure resembling that of the B subunit of *E. coli* gyrase (Fig. 28-45). Tyr 783, the residue that forms a transient phospho-Tyr covalent link with the 5'-end of a cleaved DNA strand, is located at the interface between the A' and B' subfragments of the same subunit, at the end of a narrow tunnel that opens up into the central hole. Here, the A' subfragment forms a positively charged semicircular groove that funnels into this active site tunnel. B-DNA has been modeled into this groove with a 4-nt overhang of its 5'-ending strand extending into the active site tunnel.

The dimer's two active site Tyr residues are located 27 Å apart and hence must move 35 to 40 Å toward and past each other to achieve positions that are properly staggered to link to the 5' ends of a cleaved duplex DNA. This has led to a model in which the DNA duplex to be cleaved binds in the above-described groove across the top of the heart. ATP binding to the (absent in the crystal structure) ATP-binding domain then induces a series of conformational changes in which the DNA is cleaved and the two resulting fragments, together with the two contacting B' subfragments, spread apart by at least 20 Å to permit the passage of a second DNA segment from the top of the heart through the break and into the central hole, thereby incrementing the DNA's linking number by 2. Then, in a process that is accompanied by ATP hydrolysis, the two B' subfragments come together to reseal the cleaved DNA and the DNA occupying the central hole is released from the bottom of the heart by the spreading apart of the two contacting A' subfragments. Finally, the resulting ADP and P_i are released and A' subfragments rejoin to yield recycled enzyme.

Berger, J.M., Gamblin, S.J., Harrison, S.C., and Wang, J.C., Structure and mechanism of DNA topoisomerase II, *Nature* **379,** 225–232 (1996).

(c) Aditional References

Froelich-Ammon, S.J. and Osheroff, N., Topoisomerase poisons: Harnessing the dark side of enzyme mechanisms, *J. Biol. Chem.* **270,** 21429–21432 (1995). [Discusses a variety of antibiotics and antitumor agents, some of which are widely used, that act on topoismerases to either increase the rate that they cleave DNA or inhibit the religation of their cleaved DNA, thereby causing massive DNA damage resulting in cell death.]

Sharma, A. and Mondragón, A., DNA topoisomerases, *Curr. Opin. Struct. Biol.* **5,** 39–47 (1995).

Wigley, D.B., Structure and mechanism of DNA topoisomerases, *Annu. Rev. Biophys. Biomol. Struct.* **24,** 185–208 (1995).

Yu, L., Zhu, C.-X., Tse-Dinh, Y.-C., and Fesik, S.W., A solution structure of the C-terminal single-stranded DNA-binding domain of *Eschericia coli* topoisomerase I, *Biochemistry* **34,** 7622–7628 (1995). [The NMR structure of the 14-kD C-terminal DNA-binding domain of the 97-kD *E. coli* topoisomerase I, the X-ray structure of whose 67-kD N-terminal DNA-cleaving domain is shown in Fig. 28-42 (the structure of its central zinc-binding domain remains unknown).]

6. NUCLEIC ACID SEQUENCING

(a) The X-Ray Structures of *Bam*HI and *Cfr*10I Restriction Endonucleases

The X-ray structures of two additional restriction endonucleases, *Bam*HI and *Cfr*10I, have been reported, bringing to five the number of these homodimeric enzymes of known structure. *Bam*HI makes a staggered cut in a 6-bp palindrome of sequence G↓GATCC. The X-ray structure of *Bam*HI in complex with a 12-bp duplex DNA segment containing this recognition sequence reveals that *Bam*HI binds to the major groove side of its target DNA as does *Eco*RI (Fig. 28-48; recall that *Eco*RV and *Pvu*II bind to the minor groove sides of their target DNAs). The bound B-DNA is essentially undistorted as it is in the DNA complex of *Pvu*II; it lacks the severe bends or kinks present in the DNAs bound to *Eco*RI and *Eco*RV (Figs 28-49). However, upon binding its target DNA, *Bam*HI undergoes significant conformational changes, most notably an unprecedented unwinding of its C-terminal α helix to assume an extended, partially disordered conformation (proteins often become more ordered on binding DNA).

*Cfr*10O, a restriction endonuclease from *Citrobacter freundii,* makes a staggered cut in a 6-bp DNA of sequence R↓CCGGY (where R represents purines and Y represents pyrimidines). Hence, the X-ray structure of *Cfr*10I, which was determined in the absence of DNA, represents the first of a restriction endonuclease that has a degenerate target sequence.

Despite their lack of significant sequence similarities, all five of the restriction endonucleases whose structures have been reported contain a mixed 5-stranded β sheet core sandwiched between two α helices. These proteins exhibit no other structural similarities, but in pairwise comparisons, *Bam*HI and *Eco*RI share many secondary and tertiary structural elements as do *Eco*RV and *Pvu*II, whereas *Cfr*10I has several structural elements in common with *Eco*RI and others in common with *Eco*RV.

Aggarwal, A.K., Structure and function restriction endonucleases, *Curr. Opin. Struct. Biol.* **5,** 11–19 (1995).

Bozic, D., Grazulis, S., Siksnys, V., and Huber, R., Crystal structure of *Citrobacter freundii* restriction endonuclease *Cfr*10I at 2.15 Å resolution, *J. Mol. Biol.* **255,** 176–186 (1996).

Newman, M., Strzelecka, T., Dorner, L.F., Schildkraut, I., and Aggarwal, A.K., Structure of Bam HI endonuclease bound to DNA: Partial folding and unfolding on DNA binding, *Science* **269,** 656–663 (1995).

(b) The Sequences of Several Entire Genomes Are Now Known

The base sequences of the entire genomes from three free-living organisms have recently been determined. These are, in the order they were reported, those of the bacterium *Haemophilus influenzae, Mycoplasma genitalium,* and the yeast *Saccharomyces cerevisiae.* Moreover, the genomic sequence determinations of several other microorganisms, including *E. coli,* are nearing completion. These stunning results have been made possible by advances in automated sequencing technology and, in particular, the development of computational approaches that permit the efficient assembly of tens of thousands of independent random sequences into a single sequence.

Haemophilus influenzae is a gram-negative bacterium that causes upper respiratory tract infections, ear infections, and meningitis in humans. Its 1,830,137-bp circular genome encodes 1727 protein-coding regions with, as is typical of bacteria, little (15%) noncoding DNA. Of these, 1007 (58%) have counterparts of known function in other organisms, which should help identify the functions of many of these genes.

Mycoplasma genitalium, a parasite but not necessarily a pathogen of the human genital and respiratory tracts, has a 580,070-bp circular genome, the smallest known genome of a free-living organism. It has 470 predicted protein-coding regions with 12% noncoding regions. Thus, the reduction in genome size of the *Mycoplasma* genome relative to that of *Haemophilus* has not occurred with an increase in gene density or a decrease in average gene size (900 bp in *Haemophilus* and 1040 bp in *Mycoplasma).*

The sequences of all of *S. cerevisae*'s 16 chromosomes, comprising 12.06 million bp, have as of this writing (May, 1996), been announced but not published (although these sequences can be accessed on the World Wide Web via http://genome-www.stanford.edu/Saccharomyces/). Hence, we also know the sequence of an entire eukaryotic genome, albeit one that is 1/250th the size of the 2.9 billion-bp human genome. Nevertheless, more than half its ~6000 genes have quite similar human counterparts. In an equally impressive feat, about one-fifth of the 100 million-bp genome of the nematode *Caenorhabditis elegans* had been sequenced by mid-1995.

DNA sequence data bases presently contain ~500 million bp, a number that is increasing rapidly. Thus, it appears that even without significant advances in sequencing technology (an unlikely scenario), the Human Genome Project will succeed in sequencing the entire human genome by the year 2005, if not well before.

Fleischman, R.D. et al., Whole-genome random sequencing and assembly of *Haemophilus influenzae* Rd, *Science* **269,** 496–512 (1995).

Fraser, C.M., et al., The minimal gene complement of mycoplasma genitalium, *Science* **270,** 397–403 (1995).

Hodgkin, J., Plasterk, R.H.A., and Waterston, R.H., The nematode *Caenorhabditis elegans* and its genome, *Science* **270,** 410–414 (1995).

Oliver, S.G., From DNA sequences to biological function, *Nature* **379,** 597–600 (1996).

(c) Additional References

Voet, D. and Voet, J.G., *KINEMAGES to Accompany Biochemistry, 2/E,* Wiley (1996). [Exercise 19 contains kinemages showing the structures of *Eco*RI and *Eco*RV restriction endonucleases in their complexes with their target DNAs.]

7. CHEMICAL SYNTHESIS OF OLIGONUCLEOTIDES

Eaton, B.E. and Pieken, W.A., Ribonucleosides and RNA, *Annu. Rev. Biochem.* **64,** 837–863 (1995). [Describes methods for the synthesis of modified ribonucleosides and RNA.]

Nielsen. P.E., DNA analogues with nonphosphodiester backbones, *Annu. Rev. Biophys. Biomol. Struct.* **24,** 167–183 (1995).

8. MOLECULAR CLONING

(a) Revival of Bacterial Spores from 25- to 40-Million-Year-Old Amber

Claims to having isolated ancient DNAs from fossilized organisms have been challenged on the grounds that DNA decays too fast, even in well-preserved fossils, to survive as long as one million years and that the ability of the PCR technique to amplify even one molecule of DNA makes it extremely difficult to completely exclude contaminating DNAs, particularly those from humans, from preparations purporting to be ''fossil'' DNA (see Section 28-8b below). However, bacterial spores contain proteins that specifically protect DNA, in part by stabilizing it in the A form (Section 28-2B). Thus, the DNA in bacterial spores that have been preserved in amber might be expected to survive many millions of years.

In fact, a bacterial spore taken from the stomach contents of an extinct bee that had been preserved in 25- to 40-million-year-old Dominican amber was isolated under rigorous aseptic conditions, revived, and cultured. The resulting bacteria are morphologically identical to *Bacillus sphaericus,* a symbiont of the modern descendents of the amber-entombed bee. More convincingly, the sequence of a 530-bp fragment of the 16S ribosomal RNA gene from these putative ancient bacteria more closely matches that of modern *B. sphaericus* than any of the other ~50 known bacterial rDNA sequences. Furthermore, the sequence differences between the putative ancient and modern *B. sphaericus* rDNAs tend to preserve the established secondary structure of the corresponding ribosomal RNA segment (e.g, Figs. 30-28 and 30-36). Thus, it appears likely that these *B. sphaericus*-like bacteria are indeed of ancient origin.

Cano, R.J. and Borucki, M.K., Revival and Identification of bacterial spores in 25- to 40-million-year-old Dominican amber, *Science* **268,** 1060–1064 (1995).

(b) Dinosaur DNA?

The first paper listed below presents the fragmentary sequences from a gene encoding mitochondrial cytochrome *b* that is purported to have been cloned from 80-million-year-old dinosaur bone fragments that were found in the roof of an underground coal mine. In the second paper, which is a series of technical comments on the first paper, three groups independently argue that phylogenetic analyses indicate that the ''dinosaur'' sequences are more closely related to human sequences than to those of birds and reptiles, the modern descendents of dinosaurs, and hence that these sequences are probably derived from contaminating human DNA. In a response to these contentions, the first author of the original paper argues that the poor shape and small size (174 bp) of the ''dinosaur'' DNA renders its phylogenetic analysis statistically invalid.

Woodward, S.R., Weyand, N.J., and Bunnell, M., DNA sequence from Cretaceous period bone fragments, *Science* **266,** 1229–1232 (1994).

Hedges, S.B. and Schweitzer, M.H.; Allard, M.W., Young, D., and Huyen, Y.; Zischler, H., Höss, M., Handt, O., von Haeseler, A., van der Kuyl, A.C., Goudsmit, J., and Pääbo, S.; *and* Woodward, S.R., Detecting dinosaur DNA, *Science* **268,** 1191–1194 (1995).

(c) Additional References

Anderson, W.F., Gene therapy, *Sci. Amer.* **273**(3): 124–128 (1995).

Blaese, R.M., et al., T Lymphocyte-directed gene therapy for ADA⁻ SCID: Initial trial results after 4 years, *Science* **270,** 475–480 (1995).

Crystal, R.G., Transfer of genes to humans: Early lessons and obstacles to success, *Science* **270,** 404–410 (1995). [Reviews the progress made in developing effective human gene therapy.]

Hudson, K.L., Rothenberg, K.H., Andrews, L.B., Kahn, M.J.E., and Collins, F.S., Genetic discrimination and health insurance: An urgent need for reform, *Science* **270,** 391–393 (1995).

Chapter 29

TRANSCRIPTION

2. RNA POLYMERASE

(a) The Thumb Domain of RNA Polymerase Wraps Around its DNA Template

Electron crystallographic studies of *E. coli* core RNA polymerase (which lacks the σ subunit of the RNA polymerase holoenzyme) reveals that it has undergone a dramatic conformational change relative to the holoenzyme (Fig. 29-9*a*), such that the core enzyme resembles the previously determined structure of yeast RNA polymerase II. All three of these structures exhibit a thumb-like projection, as do all other RNA and DNA polymerases of known structure. However, whereas the holoenzyme's thumb forms one wall of a ~25-Å-wide groove, that of the core enzyme curls around to contact the opposite wall of the channel, thereby forming a closed ring similar to that seen in the structure of yeast RNA polymerase II (left center of Fig. 29-16*b*). Since both the *E. coli* core enzyme and yeast RNA polymerase II catalyze the processive elongation of RNA chains (i.e., they remain bound to the DNA between each cycle of nucleotide addition) but are incapable of promoter binding without additional factors, it is proposed that the ring-like structure is indicative of an "elongation" conformation, whereas the open groove is characteristic of a "promoter-binding" conformation. The closed ring in the "elongation" complex may lock the RNA polymerase onto the DNA it is transcribing in the same way that the ring-like β subunit of *E. coli* DNA polymerase III (Fig. 31-14) is thought to lock DNA polymerase III onto the DNA that it processively replicates.

The forgoing hypothesis is supported by the X-ray structure of a chimeric T7/T3 RNA polymerase (T7: residues 1-673; T3: residues 674-884; T3 is an *E. coli* bacteriophage related to T7). Whereas the thumb in the X-ray structure of T7 RNA polymerase is well ordered and extended (Fig. 29-9*b*), that half of the T7/T3 chimera's thumb which projects furthest from the body of the polymerase is disordered. The T7/T3 chimera differs from the T7 enzyme by only 26 mostly conservative residue substitutions that are distant from the thumb domain but which cause the two proteins to crystallize in different arrangements. It is therefore likely that the differences in their thumb conformations are caused by differences in crystal packing contacts. This, in turn, suggests that the thumb is flexible, as it would have to be to wrap around the DNA being transcribed. This accounts for the observation that the mutagenic deletions of portions of the thumb from T7 RNA polymerase yields enzymes whose elongation complexes readily dissociate (i.e., have reduced processivity), but not before they have synthesized the 9-nt RNA that the native enzyme must make before it switches from "abortive" to processive RNA synthesis (changes from the closed to the open complex). It is thought that at this point, the thumb wraps around the template DNA.

Polyakov, A., Severinova, E., and Darst, S.A., Three-dimensional structure of E. coli core RNA polymerase: Promoter binding and elongation conformations of the enzyme, *Cell* **83,** 365–373 (1995).

Sousa, R., Rose, J., and Wang, B.C., The thumb's knuckle: Flexibility in the thumb subdomain of T7 RNA polymerase is revealed by the structure of a chimeric T7/T3 RNA polymerase, *J. Mol. Biol.* **244,** 6–12 (1994).

Bonner, G., Lafer, E.M., and Sousa, R., The thumb subdomain of T7 RNA polymerase functions to stabilize the ternary complex during processive transcription, *J. Biol. Chem.* **269,** 25129–25136 (1994).

(b) Additional References

Moss, T. and Stefanovsky, V.Y, Promotion and regulation of ribosomal DNA transcription in eukaryotes by RNA polymerase I, *Prog. Nucleic Acid Res. Mol. Biol.* **50,** 25–65 (1995).

Richardson, J.P., Structural organization of transcription termination factor rho, *J. Biol. Chem.* **271,** 1251–1254 (1996). [The sequence similarity of rho to the α and β subunits of the F_1-ATPase (Section 20-3C) supports the electron microscopic evidence indicating that rho consists of six identical subunits arranged in a 6-fold symmetric ring.]

3. CONTROL OF TRANSCRIPTION IN PROKARYOTES

(a) The X-Ray Structures of the *lac* Repressor and its Complexes with DNA and with IPTG

The *E. coli lac* repressor, the homotetrameric product of the *lacI* gene, has been a paradigm for gene regulation for over 35 years. In the absence of inducer (physiologically 1,6-allolactose; p. 916), it binds to three nearly palindromic operator sites in the *lac* operon: the primary operator, O_1 (Fig. 29-19), which overlaps the *lac* operon's promoter site (Figs. 29-3 and 29-20); and two auxilliary operators, O_2, which is centered 401 bp downstream of the transcription start site, within the *lacZ* (β-galactosidase) gene; and O_3, which is centered 92 bp upstream of the transcription start site, overlapping the CAP–cAMP binding site. Repressor bound to O_1 prevents RNA polymerase from completing transcriptional initiation of the *lac* operon. The simultaneous binding of a repressor tetramer to O_1 and either O_2 or O_3 requires that the intervening DNA form a loop, an arrangement that increases the efficiency of repression (see below). Upon binding inducer, the repressor changes conformation such that it no longer binds to its target operators. Then, if the CAP–cAMP complex (Fig. 29-22) is bound to its target site upstream of the *lac* promoter (Fig. 29-20), RNA polymerase binds to this promoter and transcribes of the *lac* operon.

The X-ray structure of the *lac* repressor has been determined, alone, in its complex with the nonhydrolyzable inducer isopropylthiogalactoside (IPTG; p. 917), and in its complex with a 21-bp duplex DNA segment whose sequence is a palindrome of the left half of O_1 (Fig. 29-20). The *lac* repressor, a tetramer of identical 360-residue subunits, has an unusual quaternary structure. Whereas nearly all homotetrameric proteins of known structure have D_2 symmetry (3 mutually perpendicular 2-fold axes; Fig. 7-58*b*), *lac* repressor is a V-shaped protein that has only 2-fold symmetry. Each leg of the V consists of a locally symmetric dimer of closely associated repressor subunits. Two such dimers associate rather tenuously, but with 2-fold symmetry, at the base (point) of the V to form a dimer of dimers.

Each repressor subunit consists of four functional units: an N-terminal DNA-binding domain (residues 1-45), a hinge region (residues 46-62), a sugar-binding domain (residues 63-339), and a C-terminal helix (residues 340-360; the first two units are collectively referred to as the "headpiece" because they are readily proteolytically cleaved away from the remaining still tetrameric "core" protein). The DNA-binding domain forms a compact globule containing three helices, the first two of which form a helix–turn–helix (HTH) motif. In the structures of *lac* repressor alone and that of its IPTG complex, the DNA-binding domain is not visible, apparently because the hinge region that loosely tethers it to the rest of the protein is disordered. However, in the DNA complex, in which one DNA duplex binds to each of the two dimers forming the repressor tetramer, the two DNA-binding domains extending from each repressor dimer (at the top of each leg of the V) bind in successive major grooves of a DNA molecule via their HTH motifs, much as is seen, for example, in the complexes of 434 phage repressor and *trp* repressor with their target DNAs (Figs. 29-23 and 29-25). The binding of the *lac* repressor distorts the operator DNA such that it bends away from the DNA-binding domain with a ~60-Å radius of curvature due to a ~45° kink at the center of the operator that widens the DNA's minor groove to over 11 Å and reduces its depth to less than 1 Å. These distortions permit the now ordered hinge region's single helix to bind in the minor groove so as to contact the identically-bound hinge helix from the other subunit of the same dimer. NMR measurements indicate that the DNA-binding domain, when cleaved form the repressor, binds to an operator half site without distorting the DNA. Thus, the binding of the two hinge helices to the full operator appears necessary for DNA distortion. The two DNA duplexes that are bound to each repressor tetramer are ~25 Å apart and do not interact.

The sugar-binding domain consists of two topologically similar subdomains that are bridged by three polypeptide segments. The two sugar-binding domains of a dimer make extensive contacts. IPTG binds to each sugar-binding domain between its subdomains. This does not significantly change the conformations of these subdomains, but it changes the angle between them. Although the hinge region is not visible in the IPTG complex, model building indicates that, since the dimer's two hinge helices extend from its sugar-binding domains, this conformational change levers apart these hinge helices by 3.5 Å such that they and their attached HTH motifs can no longer simultaneously bind to their operator half-sites. Thus, inducer binding, which is allosteric within the dimer (has a positive homotropic effect; Section 9-4B), greatly loosens the repressor's grip on the operator.

The C-terminal helices from each subunit, which are located on the opposite end of each subunit from the DNA-binding portion (at the point of the V), associate to form a 4-helix barrel that holds together the two repressor dimers, thereby forming the tetramer. The allosteric effects of inducer binding within each dimer are apparently not transmitted between dimers. Moreover, the *E. coli* **purine repressor (PurR),** which is homologous to the *lac* repressor but lacks its C-terminal helix, crystallizes as a dimer whose X-ray structure closely resembles that of the *lac* repressor dimer. What then is the function of *lac* repressor tetramerization?

Model building suggests that when the *lac* repressor tetramer simultaneously binds to both the O_1 and O_3 operators, the 93-bp DNA segment containing them forms a ~80-Å wide loop. Furthermore, the CAP–cAMP binding site is exposed on the inner surface of the loop. Adding the CAP–cAMP at its proper position to this model reveals that the ~90° curvature which CAP–cAMP binding imposes on DNA (Fig. 29-22) has the correct direction and magnitude to stabilize the DNA loop, thereby stabilizing this putative CAP–cAMP–*lac* repressor–DNA complex. It may seem paradoxical that the binding of CAP–cAMP, a transcriptional activator, stabilizes the repressor–DNA complex. However, when both glucose and lactose are in short supply, it is important that the bacterium lower its basal rate of *lac* operon expression in order to conserve energy. The binding site (promoter) for RNA polymerase is also located on the inner surface of the loop. Thus, the large size of the RNA polymerase molecule would prevent it from fully engaging the promoter in this looped complex, thereby maximizing repression.

Lewis, M., Chang, G., Horton, N.C., Kercher, M.A., Pace, H.C., Schumacher, M.A., Brennan, R.G., and Lu, P., Crystal structure of the lactose operon repressor and its complexes with DNA and inducer, *Science* **271**, 1247–1254 (1996).

Schumacher, M.A., Choi, K.Y., Zalkin, H., and Brennan, R.G., Crystal structure of lacI member, PurR, bound by DNA: minor groove binding by a helices, *Science* **266**, 763–770 (1994); Schumacher, M.A., Choi, K.Y., Lu, F., Zalkin, H., and Brennan, R.G., Mechanism of corepressor-mediated specific DNA binding by the purine repressor, *Cell* **83**, 147–155 (1995); *and* Nagadoi, A., et al., Structural comparison of the free and DNA-bound forms of the purine repressor DNA-binding domain, *Structure* **3**, 1217–1224 (1995). [X-ray and NMR studies of the structure and DNA-binding properties of the *E. coli* purine repressor, a dimeric homolog of the *lac* repressor.]

Friedman, A.M., Fischmann, T.O., and Steitz, T.A., Crystal structure of *lac* repressor core tetramer and its implications for DNA looping, *Science* **268**, 1721–1727 (1995). [Shows a similar structure to that of the core tetramer in the X-ray structures of the intact *lac* repressor.]

(b) The α Subunit of RNA Polymerase Binds Upstream Promoter Elements and Activator Proteins

For most *E. coli* structural genes, there is considerable correlation between promoter efficiency (the rate that transcription is initiated) and the similarity of their –10 and –35 elements to their respective consensus sequences (Fig. 29-10). However, many genes, particularly those with highly efficient promoters, contain additional promoter elements that are located upstream of the –35 region. For example, the seven *E. coli* *rrn* genes, which encode all ribosomal RNAs (Section 29-4), are responsible for over 60% of the RNA synthesis in rapidly growing cells. All of the *rrn* genes have an A+T-rich segment immediately upstream of their –35 site (as can be seen for the three *rrn* genes in Fig. 29-10). Deletion of this **upstream (UP) element** reveals that it stimulates transcription by a factor of 30 *in vivo* as well as *in vitro* in the absence of proteins other than RNA polymerase. Moreover, when fused to other promoters, the UP element also stimulates transcription indicating that it is an independent module.

The –10 and –35 regions in promoters associated with most genes are recognized by the σ^{70} subunit of RNA polymerase (recall that *E. coli* RNA polymerase holoenzyme has the subunit composition $\alpha_2\beta\beta'\sigma$, in which different σ subunits recognize the promoters for different classes of proteins). However, mutations in the C-terminal domain of RNA polymerase's α subunit (αCTD) eliminate transcriptional stimulation by the UP element, although these mutant RNA polymerases effectively initiate transcription at promoters that lack UP elements. Moreover, wild-type but not the mutant αCTDs protect the UP element in footprinting experiments. Thus, the αCTD must directly recognize the UP element.

Many *E. coli* promoters are, in part, regulated by one or more activator proteins such as the CAP–cAMP complex. Deletion or certain mutations of the αCTD abolishes this effect, which suggests that the αCTD is also the target of these activator proteins.

The NMR structure of the αCTD (residues 233-329 of the α subunit) reveals a compact structure containing 4 helices and 2 long arms enclosing a hydrophobic core. The residues that NMR experiments have implicated in binding the UP element and the CAP-cAMP complex are all located on the same surface region of αCTD.

Busby, S. and Ebright, R.H., Promoter structure, promoter recognition, and transcription activation in prokaryotes, *Cell* **79,** 743–746 (1994).

Jeon, Y.H., Negishi, T., Shirakawa, M., Yamazaki, T., Fujita, N., Ishihama, A., and Kyogoku, Y., Solution structure of the activator contact domain of the RNA polymerase α subunit, *Science* **270,** 1495–1499 (1995).

Ross, W., Gosink, K.K., Salomon, J., Igarashi, K., Zou, C., Ishihama, A., Severinov, K., and Gourse, R.L., A third recognition element in bacterial promoters: DNA binding by the α subunit of RNA polymerase, *Science* **262,** 1407–1413 (1993).

(c) Structure of the *trp* RNA-Binding Attenuation Protein (TRAP) From *Bacillus subtilis*

In *E. coli* and several other enteric bacteria, tryptophan biosynthesis is regulated, in part, by a mechanism known as attenuation, in which a ribosome binds to the nascent mRNA for the *trp* operon and commences synthesizing its so-called leader peptide. The mRNA segment encoding the leader peptide has two mutually exclusive secondary structures (Fig. 29-30): a "terminator" (a hairpin followed by several U's; Fig. 29-15), which causes RNA polymerase to terminate the transcription of the succeeding portion of the *trp* operon, and an "antiterminator", which by preventing terminator formation, permits transcription to continue. The initial segment of the leader peptide contains the rare dipeptide Trp-Trp. When tryptophan is abundant, the ribosome rapidly translates this newly synthesized mRNA segment and then translates the adjacent antiterminator-containing mRNA segment. Ribosome binding prevents the formation of the antiterminator structure, thereby permitting the formation of the subsequently synthesized terminator structure (Fig. 29-31*a*). However, when tryptophan is scarce, the ribosome stalls at the codons specifying the Trp-Trp dipeptide, thus permitting the antiterminator to form, resulting in the further transcription of the complete *trp* operon mRNA (Fig. 29-31*b*).

Tryptophan biosynthesis in *Bacillus subtilis* is also regulated by attenuation but via a mechanism in which the ribosome is, in effect, replaced by a tryptophan-binding protein known as *trp* RNA-binding attenuation protein (TRAP). The TRAP–tryptophan complex specifically binds to the mRNA segment that encodes the leader peptide so as to prevent the formation of the antiterminator and hence permit the formation of the terminator. In the absence of tryptophan, TRAP does not bind to the mRNA, thereby permitting the antiterminator to form and the *trp* operon to be transcribed.

The X-ray structure of TRAP in complex with tryptophan reveals a remarkable protein. It consists of 11 identical 75-residue subunits symmetrically arranged in a ring such that the protein has the shape of an 85-Å in diameter doughnut that has a 23-Å in diameter hole. Each subunit has a novel fold consisting of a sandwich of a 3-stranded and a 4-stranded antiparallel β sheet whose strands extend more or less radially and which superficially resembles the immunoglobulin fold (Fig. 34-24). However, the 4-stranded sheet in each subunit is continued by the 3-stranded sheet in an adjoining subunit so that TRAP consists of eleven overlapping 7-

stranded antiparallel β sheets arranged in a ring. Eleven tryptophan molecules are bound to TRAP, one between each pair of its adjacent 7-stranded β sheets, such that each tryptophan is completely enclosed by protein. Consequently, in the absence of tryptophan, TRAP must have a different conformation to provide tryptophan access to its binding site.

TRAP's RNA recognition sequence, which is located between nucleotides 36 and 91 of the *trp* leader mRNA, contains 7 GAG and 4 UAG trinucleotides, each separated by 2 or 3 nucleotides. Sequential mutatgenic deletions of these sites from either end of the mRNA causes progressively decreasing TRAP activity. This suggests that, in the presence of tryptophan, the leader mRNA segment wraps around TRAP such that a GAG or UAG trinucleotide is bound to each subunit. This permits nucleotides 108 to 136 to form a terminator structure. In the absence of tryptophan, TRAP does not bind to the leader mRNA. This permits nucleotides 59 to 111 to form an antiterminator, a hairpin that prevents the formation of the terminator hairpin.

Antson, A.., Otridge, J., Brzozowski, A.M., Dodson, E.J., Dodson, G.G., Wilson, K.S., Smith, T.M., Yang, M., Kurecki, T., and Gollnick, P., The structure of the trp RNA-binding attenuation protein, *Nature* **374,** 693–700 (1995).

(d) Additional References

Nelson, H.C.M., Structure and function of DNA-binding proteins, *Curr. Opin. Genet. Dev.* **5,** 180–189 (1995).

Voet, D. and Voet, J.G., *KINEMAGES to Accompany Biochemistry, 2/E,* Wiley (1996). [Exercise 20 consists of kinemages of 434 repressor and 434 Cro protein in their complexes with a 20-bp DNA containing their target sequence.]

4. POST-TRANSCRIPTIONAL PROCESSING

(a) X-Ray Structure of an RNA-Binding Domain of U1A Spiceosomal Protein in Complex with an RNA Hairpin

The RNA-recognition motif (RRM), which is also known as the **ribonucleoprotein (RNP) motif,** is the most common RNA-binding motif. It has been identified in over 200 different RNA-binding proteins, including 5 of the 7 ribosomal proteins of known structure (Section 30-3A). **U1A protein,** a component of the **U1-snRNP,** contains two RRMs separated by a ~100-residue polypeptide segment.

The X-ray structure has been determined of the 98-residue N-terminal RRM of U1A in complex with a 21-nt RNA that differs in sequence from hairpin II of U1 snRNA by only three nucleotides at its 5'- and 3'-termini. The hairpin has a 5-bp stem, a 10-nt loop of sequence AUUGCACUCC, and a 1-nt overhang at its 3' end. The structure of the protein in the complex resembles those diagrammed in Fig. 30-31. Its only significant differences with the previously determined structure of the uncomplexed protein are small movements of two loops and an ordering of its previously disordered C-terminal 13 residues. The RNA's 10-nt loop forms an open structure that extends from the double helical stem in a direction that is nearly perpendicular to its helix axis and with all of the loop's bases splayed out from its center. The AUUGCAC segment of the loop fits into a groove on the surface of the protein where its sugar-phosphate backbone interacts with several positively charged side chains and with each base stacked on either another base, an aromatic amino acid side chain, or both. These 7 nucleotides participate in an extensive network of direct and water-mediated hydrogen bonds involving the protein side chains and main chain. The remaining UUC segment of the loop as well as the 5-bp stem extend into solution out of contact with the protein.

Oubridge, C., Ito, N., Evans, P.R., Teo, C.-H., and Nagai, K., Crystal structure at 1.92 Å resolution of the RNA-binding domain of the U1A spliceosomal protein complexed with an RNA hairpin, *Nature* **372,** 432–438 (1994).

Nagai, K., Oubridge, C., Ito, N., Avis, J., and Evans, P., The RNP domain: a sequence-specific RNA-binding domain involved in processing and transport of RNA, *Trends Biochem. Sci.* **20,** 235–240 (1995).

(b) X-Ray Structure of an All-RNA Hammerhead Ribozyme

Hammerhead ribozymes, the simplest known ribozymes, catalyze self-cleavage reactions. The hammerhead ribozyme whose X-ray structure is shown in Fig. 29-44 contained a DNA pseudosubstrate strand to prevent its cleavage and was crystallized from a solution containing high concentrations of a monovalent ions [$2M$ Li_2SO_4 or $(NH_4)_2SO_4$]. This raised two questions: Does the DNA strand significantly perturb the structure of the ribozyme, and since a divalent metal ion such as Mg^{2+} is thought necessary to stabilize the ribozyme's trigonal bipyrimidal reaction intermediate, is the the high concentration of monovalent ions responsible for the apparent absence of divalent metal ions in the vicinity of the cleavage site?

To answer these questions, an all-RNA hammerhead ribozyme, which differs in sequence and connectivity from that in Fig. 29-44*a*, was synthesized in which the C at the cleavage site (position 17) was replaced with 2'-methoxy-C, thereby blocking the formation of the 2',3'-cyclic phosphodiester reaction product at this position. This RNA, which was crystallized from a solution that contained 10 mM Mg^{2+} and relatively low concentrations of monovalent cations in a different, assumed a different crystal form than that of the DNA-containing ribozyme. Nevertheless, with the exception of some additional stabilizing hydrogen bonds in the all-RNA structure, these two molecules have quite similar conformations suggesting that both closely resemble the solution structure of a functional hammerhead ribozyme.

Since Mg^{2+} ions and H_2O molecules both contain 10 electrons, the only way that the two may be differentiated in an electron density map of a macromolecule (which is of too low a resolution to visualize H atoms) is by the characteristic arrangement of ligands about the Mg^{2+} ion. Nevertheless, five potential Mg^{2+} ions were identified among the numerous solvent peaks visible in the all-RNA structure. One of them is located in the vicinity of C-17 at the cleavage site but is just out of striking distance from the reaction site to participate in catalysis. However, this Mg^{2+} ion is also associated with the C residue in the CUGA sequence at positions 3 to 6 (Fig. 29-44*a*) that is absolutely conserved in all hammerhead ribozymes. In the various X-ray structures of tRNA[Phe], the Cm-U-Gm-A segment in its anticodon loop (Fig. 30-14; Cm and Gm have methoxy groups at their 2' positions) takes up a nearly identical conformation to that of the ribozymal CUGA segments but binds divalent metal ions at somewhat different positions. This has led to the formulation of a speculative mechanism in which the Mg^{2+} ion moves from its observed position to one similar to that seen in a tRNA[Phe] structure. There it can stabilize an arrangement that permits the nucleophilic attack by the 2'-hydroxyl of C-17 on the phosphate group substituent to C3' in a direction that is in line with the P—O5' bond to the leaving group (as in Fig. 15-7*b*). The Mg^{2+} ion may activate this nucleophilic attack through an Mg^{2+}-polarized hydroxyl group that abstracts the proton from the attacking 2'-hydroxyl group (Section 14-1C).

Scott, W.G., Finch, J.T., and Klug, A., The crystal structure of an all-RNA hammerhead ribozyme: A proposed mechanism for RNA catalytic cleavage, *Cell* **81,** 991-1002 (1995).

(c) Additional References

Ares, M., Jr. and Weiser, B., Rearrangement of snRNA structure during assembly and function of the spliceosome, *Prog. Nucleic Acid Res. Mol. Biol.* **50,** 131–159 (1995).

Manley, J.L., A complex protein assembly catalyzes polyadenylation of mRNA precursors, *Curr. Opin. Genet. Dev.* **5,** 222–228 (1995); *and* Keller, W., No end yet to messenger RNA 3' processing, *Cell* **81,** 829–832 (1995).

Maxwell, E.S. and Fournier, M.J., The small nucleolar RNAs, *Annu. Rev. Biochem.* **35,** 897–934 (1995). [Discusses **small nucleolar RNAs (snoRNAs),** which in eukaryotes, are implicated in the processing of ribosomal RNAs.]

Michel, F. and Ferat, J.-L., Structure and activity of group II introns, *Annu. Rev. Biochem.* **64,** 35–61 (1995).

Nagai, K., RNA-protein complexes, *Curr. Opin. Struct. Biol.* **6,** 53–61 (1996).

Nagai, K. and Mattaj, I.W. (Eds.), *RNA-Protein Interaction,* IRL Press (1994).

Nierlich, D.P. and Murakawa, G.J., The decay of bacterial messenger RNA, *Prog. Nucleic Acid Res. Mol. Biol.* **52,** 153–216 (1996).

Scott, J., A place in the world for RNA editing, *Cell* **81,** 833–836 (1995). [Reviews **substitution editing** in mRNA, the process whereby one base is enzymatically converted to another, thereby changing the genetic message, as occurs in the conversion of apoB100 mRNA to apoB48 mRNA.]

Shen, L.X. and Tinoco, I., Jr., The structure of an RNA pseudoknot that causes efficient frameshifting in mouse mammary tumor virus, *J. Mol. Biol.* **247,** 963–978 (1995).

Shuman, S., Capping enzyme in eukaryotic mRNA synthesis, *Prog. Nucleic Acid Res. Mol. Biol.* **50,** 101–129 (1995).

Simpson, L. and Thiemann, O.H., Sense from nonsense: RNA editing in mitochondria of kinetoplastid protozoa and slime molds, *Cell* **81,** 837–840 (1995).

Tuschl, T., Thomson, J.B., and Eckstein, F., RNA cleavage by small catalytic RNAs, *Curr. Opin. Struct. Biol.* **5,** 296–302 (1995).

Chapter 30

TRANSLATION

1. THE GENETIC CODE

Cornish, V.W., Mendel, D., and Schulz, P., Probing protein structure with an expanded genetic code, *Angew. Chem. Int. Ed. Engl.* **34,** 621–633 (1995). [Describes a method of incorporating chemically synthesized amino acid residues into specific sites in proteins via a ribosomal system and how these unnatural residues affect the properties of the proteins.]

Tate, W.P., Poole, E.S., and Mannerling, S.A., Hidden infidelities of the translational stop signal, *Prog. Nucleic Acid Res. Mol. Biol.* **52,** 293–335 (1996).

2. TRANSFER RNA

(a) A Bonanza of Aminoacyl-tRNA Synthetase Structures

In 1995, the X-ray structures of 6 species of aminoacyl-tRNA synthetases (RSs) were reported, thereby more than doubling the 5 such structures that were previously known. Two of them, MetRS and TrpRS, are Class I RSs, and the remainder, GlyRS, HisRS, LysRS and PheRS, are Class II RSs. The various structures in each class show both clear similarities and large differences. Thus, in the Class II RSs, AspRS, LysRS, and SerRS have sufficient structural similarities in their catalytic domains that they have been placed in a separate subclass as were GlyRS and HisRS, which also have homologous C-terminal domains. In PheRS, an $\alpha_2\beta_2$ heterotetramer and the only RS of known structure with two different subunits (all others are either monomers or homodimers), the catalytic domain of the α subunit and ''catalytic-like'' domain of the β subunit are topologically identical.

In the Class I RSs, TrpRS and Tyr RS are far more structurally similar than might be expected on the basis of their sequence alignments. The catalytic domains of GluRS and GlnRS likewise exhibit considerable structural similarity although their sequences are also similar (many bacteria lack a GlnRS; their tRNA^{Gln} and tRNA^{Glu} are both aminoacylated by GluRS yielding Glu–tRNA^{Gln} and Glu–tRNA^{Glu} with the former being subsequently converted to Gln–tRNA^{Gln} by a transamidase).

The structures of HisRS and TrpRS were determined in complex with histidyl-adenylate and tryptophanyl-adenylate, both of which were formed, *in situ,* by soaking crystals of His RS and Trp RS with ATP together with histidine or tryptophan, respectively. Thus, these crystalline RSs have at least partial enymatic activity. The LysRS structure contains a bound lysine, the only known structure of an RS in complex with its target amino acid.

Cusack, S., Eleven down and nine to go, *Nature Struct. Biol.* **2,** 824–831 (1995). [Reviews and compares the structures of the 11 aminoacyl-tRNA synthetases whose X-ray structures are presently known.]

Arnez, J.G., Harris, D.C., Mitschler, A., Rees, B., Franklyn, C.S., and Moras, D., Crystal structure of histidyl-tRNA synthetase from *Eschericia coli* complexed with histidyl-adenylate, *EMBO J.* **14,** 4143–4155 (1995).

Belrhali, H., Yaremchuk, A., Tukalo, M., Berthet-Colominas, C., Rasmussen, B., Büsecke, P., Diat, O., and Cusack, S., The structural basis for seryl-adenylate and Ap$_4$A synthesis by seryl-tRNA synthetase, *Structure* **3,** 341–352 (1995).

Doublié, S., Bricogne, G., Gilmore, C., and Carter, C.W., Jr., Tryptophanyl-tRNA synthetase crystal structure reveals an unexpected homology to tyrosyl-tRNA synthetase, Structure **3,** 17–31 (1995).

Logan, D.T., Mazuaric, M.-H., Kern, D., and Moras, D., Crystal structure of glycyl-tRNA synthetase from *Thermus thermophilus,* *EMBO J.* **14,** 4156–4167 (1995).

Mosyak, L., Reshetnikova, L., Goldgur, Y., Delarue, M., and Safro, M.G., Structure of phenylalanyl-tRNA synthetase from *Thermus thermophilus, Nature Struct. Biol.* **2,** 537–547 (1995).

Nurecki, O., Vassylyev, D.G., Katayanagi, K., Shimizu, T., Sekine, S., Kigawa, T., Miyazawa, T., Yokoyama, S., and Morikawa, K., Architectures of class-defining and specific domains of glutamyl-tRNA synthetase, *Science* **267,** 1958–1965 (1995).

Onesti, S., Miller, A.D., and Brick, P., The crystal structure of lysyl-tRNA synthetase (LysU) from *Eschericia coli, Structure* **3,** 163–176 (1995).

(b) Additional References

Draper, D.E., Protein-RNA recognition, *Annu. Rev. Biochem.* **64,** 593–620 (1995).

Gabriel, K., Schneider, J., and McClain, W.H., Functional evidence for indirect recognition of G·U in tRNA[Ala] by alanyl-tRNA synthetase, *Science* **271,** 195–97 (1996). [The observation that tRNA[Ala] with the non-Watson-Crick base pairs G·U (wild-type), C·A, or G·A at its nucleotide positions 3 and 70 were similarly charged by AlaRS, whereas tRNA[Ala] with G·C in this position is inactive suggests that AlaRS senses the distortion of the acceptor stem at these identity element positions rather than only binding the functional groups of G·U.]

Hoagland, M., Biochemistry or molecular biology? The discovery of 'soluble RNA', *Trends Biochem. Sci.* **21,** 77–80 (1996). [An eyewitness report.]

Schimmel, P. and de Pouplana, L.R., Transfer RNA: From minihelix to genetic code, *Cell* **81,** 983–986 (1996).

Voet, D. and Voet, J.G., *KINEMAGES to Accompany Biochemistry, 2/E,* Wiley (1996). [Exercise 21 consists of kinemages of yeast tRNA[Phe] and Exercise 22 consists of kinemages of *E. coli* GlnRS·tRNA[Gln]·ATP.]

3. RIBOSOMES

(a) More Detailed Images of Ribosomes Reveal Labyrinthine Particles with Bound tRNAs

Two independent studies in which *E. coli* ribosomes were embedded in vitreous ice and visualized through the analysis of numerous cryo-electron microscopy-based images, have yielded ~25-Å resolution, 3-dimensional models of the ribosome showing far more detail than was previously apparent. To a first approximation, the *E. coli* ribosome is a sphere of radius ~110 Å, which has a volume of ~5600 nm³. However, its volume based on its mass is ~2600 nm³ and hence 55% of this sphere is solvent-filled. Hence, although the general outline of the ribosome resembles that portrayed in Figs. 30-26*a* and 30-27, it has numerous newly revealed clefts and cavities as well as tunnels that penetrate both the 30S and 50S subunits, and a gap between them (it gives the impression of a lump of Swiss cheese or a particularly worm-eaten apple).

The tunnels, one of which is probably that seen in Fig. 30-27, are postulated to form the entry and exit pathways for mRNA. They are too narrow to admit more than a single strand of mRNA, thereby preventing any intrastrand base pairing interactions from interfering with the translational process. The model is of sufficiently high resolution that specific double helical segments of ribosomal RNA (Fig. 30-28) could be confidently located in several of its bulges, albeit only roughly.

Cryo-electron microscopy studies of poly(U)-programmed ribosomes revealed three previously unseen crescent-shaped objects lying in close proximity in the gap between the 30S and 50S subunits. These are apparently tRNA molecules occupying the ribosome's A, P, and E sites. In accordance with the results of previous experiments, the tRNA nearest the large subunit's L7/L12 stalk (Fig. 30-34*b*) occupies the A site, that furthest from the A site occupies the E site, and the tRNA between the other two tRNAs occupies the P site. The A- and P-site tRNAs, as expected from a consideration of the translational process, are nearly in contact at both their putative anticodon loops and CCA ends. The CCA end of the P-site tRNA points towards the entrance of a tunnel through which the newly synthesized polypeptide chain had been postulated to exit the ribosome.

Frank, J., Zhu, J., Penczek, P., Li, Y., Srivastava, S., Verschoor, A., Radermacher, M., Grassucci, R., Lata, R.K., and Agrawal, R.K., A model of protein synthesis based on cryo-electron microscopy of the *E. coli* ribosome, *Nature* **376,** 441–444 (1995); *and* Agrawal, R.K., Penczek, P., Grassucci, R.A., Li, Y., Leith, A., Nierhaus, K.H., and Frank, J., Direct visualization of A-, P-, and E-site transfer RNAs in the *Eschericia coli* ribosome, *Science* **271,** 1000–1002 (1996).

Stark, H., Mueller, F., Orlova, E.V., Schatz, M., Dube, P., Erdemir, T., Zemlin, F., Brinacombe, R., and van Heel, M., The 70S *Eschericia coli* ribosome at 23 Å resolution: fitting the ribosomal RNA, *Structure* **3,** 815–821 (1995).

(b) The X-Ray Structures of Phe-tRNA^Phe·EF-Tu·GDPNP and EF-Tu·EF-Ts

EF-Tu in complex with an aminoacyl-tRNA and GTP delivers the aminoacyl-tRNA to the A site of the ribosome in its posttranslocational state (Fig. 30-46) and, upon hydrolysis of the GTP to GDP + P$_i$, dissociates from the ribosome while facilitating its conversion to the pretranslocational state. The X-ray structure of the Phe-tRNA^{Phe}·EF-Tu·GDPNP ternary complex reveals that these two macromolecues associate to form a corkscrew-shaped complex in which the EF-Tu and the tRNA's acceptor stem form a knob-like handle and the tRNA's anticodon helix forms the screw. The conformations of the two macromolecules closely resemble those seen in the X-ray structures of EF-Tu·GDPNP (Fig. 30-45a) and uncomplexed tRNA^{Phe} (Fig. 30-14b). The macromolecules associate rather tenuously with the 3'-CCA–Phe segment of the Phe-tRNA^{Phe} binding in the cleft between domains 1 and 2 of the EF-Tu·GDPNP (red and blue in Fig. 30-45a), the 5'-phosphate of the tRNA binding in a depression at the junction of EF-Tu's three domains, and one side of the TψC stem of the tRNA making contacts with exposed main chain and side chains of EF-Tu domain 3 (green in Fig. 30-45a). Presumably, all species of tRNA make these contacts with EF-Tu.

The X-ray structure of the ternary complex suggests why EF-Tu binds all aminoacylated tRNAs except initiator tRNAs, but never uncharged tRNAs. Evidently, the tight association of the aminoacyl group with EF-Tu greatly increases the affinity of EF-Tu for the otherwise loosely bound tRNA. The first base pair of tRNA$_f^{Met}$ is mismatched (Fig. 30-40) and hence this initiator tRNA has a 3' overhang of 5 nt vs 4 nt for an elongator tRNA. This, together with the formyl group on its appended fMet residue apparently prevents fMet-tRNA$_f^{Met}$ from binding to EF-Tu.

EF-G·GTP functions in the translocation stage of the ribosomal elongation cycle (Fig. 30-44) by facilitating the conversion of the ribosome from its pretranslocational to its postranslocational state (Fig. 30-46). The binding of EF-G and EF-Tu to the ribosome is mutually exclusive because both bind to the same site. Yet, the first two domains in the X-ray structure of EF-Tu·GDP closely resemble those in EF-Tu·GDPNP rather than those in EF-Tu·GDP (Fig. 30-45). This, it has been suggested, is because the two elongation factors have reciprocal functions with EF-Tu·GTP facilitating the conversion of the ribosome from its post- to its pretranslocational state and EF-G·GTP promoting the reverse transition. This idea is supported by the intriguing observation that the Phe-tRNA^{Phe}·EF-Tu·GDPNP and EF-G·GDP complexes are almost identical in appearance: EF-G's three C-terminal domains (green, red and yellow in Fig. 30-45b), which have no counterparts in EF-Tu, closely resemble the EF-Tu-bound tRNA in shape. However, this argument is by no means conclusive because the structures of EF-Tu·GDP and EF-G·GDP, neither of which bind to the ribosome, are not necessarily informative about ribosomal properties, whereas the structure of EF-G·GTP, which does bind to the ribosome, is unknown. Whatever the case, the molecular mimicry by the three C-terminal domains of EF-G for tRNA has provided a substantial although still enigmatic clue as to the mechanism of translation.

EF-Tu has a 100-fold higher affinity for GDP than GTP. Hence, replacement of the EF-Tu-bound GDP by GTP must be facilitated by the interaction of EF-Tu with the guanine nucleotide releasing factor (GRF) known as EF-Ts (Fig. 30-44, *top*). The X-ray structure of the EF-Tu·EF-Ts complex shows that the EF-Tu has a conformation resembling that of its GDP complex (Fig. 30-45a) but with its domains 2 and 3 swung away from domain 1 by ~18°. EF-Ts is an elongated molecule that binds along the right side of EF-Tu as shown in Fig. 30-45a where it contacts EF-Tu's domains 1 and 3. The intrusive interactions of EF-Ts side chains with the GDP binding pocket on EF-Tu disrupts the Mg^{2+} ion binding site, thereby reducing the affinity of EF-Tu for GDP and facilitating its exchange for GTP.

Nissen, P., Kjeldgaard, M., Thirup, S., Polekhina, G., Reshetnikova, L., Clark, B.F.C., and

Nyborg, J., Crystal structure of the ternary complex of Phe-tRNA[Phe], EF-Tu and a GTP analog, *Science* **270,** 1464–1472 (1995); *and* Nyborg, J., Nissen, P., Kjeldgaard, Thirup, S., Polekhina, G., and, Clark, B.F.C., Structure of the ternary complex of EF-Tu: macromolecular mimicry in translation, *Trends Biochem. Sci.* **21,** 81–82 (1996).

Moore. P.B., Molecular mimicry in protein synthesis, *Science* **270,** 1453–1454 (1995).

Kawashima, T., Berthet-Columinas, C., Wulff, M., Cusack, S., and Leberman, R., The structure of the *Eschericia coli* EF-Tu·EF-Ts complex at 2.5 Å resolution, *Nature* **379,** 511–518 (1996).

Abel, K. and Jurnak, F., A complex profile of protein elongation: translating chemical energy into molecular movement, *Structure* **4,** 229–238 (1996).

(c) C74 of tRNA Forms a Watson–Crick Base Pair with G2252 of the 23S RNA

The 3' ends of all mature tRNAs have the sequence CCA, which is essential for functional interactions between the tRNA and the peptidyl transferase P site. These two conserved cytosine bases are protected from methylation by dimethyl sulphate when the tRNA is bound to the ribosomal P site. Conversely, two consecutive universally conserved G's in the 23S RNA, G2252 and G2253, are protected from chemical modification by P site-bound aminoacyl-tRNAs and/or CCA-ending oligonucleotide segments. The deletion of a tRNA's C75 but not its A76 results in specific loss of protection of these two ribosomal bases, which suggests that the tRNA's conserved cytosines base pair with the 23S RNA's conserved guanines.

This hypothesis was tested by ascertaining the effects of mutations at G2252 and G2253 on the ribosome's peptidyl transferase function. Mutations at G2252 confer a dominant lethal phenotype on *E. coli*. However, the effects of mutations at G2252 on the chemical protection of tRNA[Phe] are suppressed by compensatory mutations, in the Watson–Crick sense, at tRNA's C74. Similar effects are seen in the binding of the olignucleotide CACCA(Ac-Phe) and its variants to the ribosomal P site. This strongly suggests that C74 of tRNA and G2252 of 23S RNA interact by Watson–Crick base pairing, the first demonstration of a specific base-base interaction between tRNA and rRNA. In contrast, mutations at G2253, which only confer a recessive slow growth phenotype on *E. coli,* showed no clear pattern of suppression in chemical protection experiments and had little effect on the binding of CACCA (Ac-Phe) or its variants to the ribosomal P site. Thus, there is no clear requirement for base pairing between C75 and G2253.

Samaha, R.R.. Green, R., and Noller, H.F., A base pair between tRNA and 23S rRNA in the peptidyl transferase centre of the ribosome, *Nature* **377,** 309–314 (1995).

Lieberman, K.R. and Dahlberg, A.E., Ribosome-catalyzed peptide-bond formation, *Prog. Nucleic Acid Res. Mol. Biol.* **50,** 1–23 (1995).

(d) Additional References

Biou, V., Shu, F., and Ramakrishnan, V., X-ray crystallography shows that translational initiation factor IF3 contains two compact α/β domans linked by an α-helix, *EMBO J.,* **14,** 4056–4064 (1995).

Liljas, A. and Garber, M., Ribosomal proteins and elongation factors, *Curr. Opin. Struct. Biol.* **5,** 721–727 (1995).

Stansfield, I., Jones, K.M., and Tuite, M.F., The end is in sight: terminating translation in eukaryotes, *Trends Biochem. Sci.* **20,** 489–491 (1995).

4. CONTROL OF EUKARYOTIC TRANSLATION

Chen, J.-J. and London, I.M., Regulation of protein synthesis by heme-regulated eIF-2α kinase, *Trends Biochem. Sci.* **20,** 105–108 (1995).

Morris, D.R., Growth control of translation in mammalian cells, *Prog. Nucleic Acid Res. Mol. Biol.* **51,** 339–363 (1995).

Miller, P.S., Development of antisense and antigene oligonucleotide analogs, *Prog. Nucleic Acid Res. Mol. Biol.* **52,** 261–291 (1996).

Wagner, R.W., Gene inhibition using antisense oligodeoxynucleotides, *Nature* **372,** 333–335 (1994).

5. POST-TRANSLATIONAL MODIFICATION

(a) Protein Splicing

The discovery of **protein splicing** has provided yet another twist to the Central Dogma of molecular biology. In this phenomenon, which has been shown to occur in yeast and in several prokaryotes, a polypeptide segment named an **intein**, which is located between a so-called **N-extein** on its N-terminal side and a **C-extein** on its C-terminal side, excises itself via a not fully characterized mechanism that joins together the N-extein and the C-extein to form a functional protein – a process analogous to RNA self-splicing (Section 29-4B). The information for carrying out protein splicing resides entirely within the intein, which has a Cys or Ser at its N-terminus and His-Asn at its C-terminus. Several of the self-excised intein proteins are site-specific DNA endonucleases that catalyze the genetic mobility of their DNA coding sequences.

Cooper, A.A. and Stevens, T.H., Protein splicing: self-splicing of genetically mobile elements at the protein level, *Trends Biochem. Sci.* **20,** 351–356 (1995).

6. PROTEIN DEGRADATION

(a) X-Ray Structure of the 20S Proteasome

The 26S proteasome (Fig. 30-62) is a multisubunit eukaryotic protein that catalyzes the ATP-dependent hydrolysis of ubiquitin-linked proteins, yielding 6- to 9-residue oligopeptides that are subsequently degraded to their component amino acids by cytosolic exopeptidases. The 26S proteasome consists of a **20S proteasome,** the barrel-shaped catalytic core of the 26S proteasome, and its 19S "caps", which associate with the ends of the 20S proteasome. The ATP-dependent degradation of ubiquinated proteins is mediated by the 26S proteasome; the 20S proteasome can only hydrolyze unfolded proteins in an ATP-independent fashion. Hence, the 19S caps probably function to engulf and unfold the incoming protein substrate.

The 20S proteasome, which occurs in the nuclei and cytosol of all eukaryotic cells so far examined, has also been observed in the archaebacterium *Thermoplasma acidophilum*. The *T. acidophilum* 20S proteosome consists of 14 copies each of α (233 residues) and β (203 residues) subunits that electron microscopy studies reveal form a 150-Å long × 110-Å in diameter barrel in which the subunits are arranged in 4 stacked rings (as is evident in the central portion of the 26S proteasome seen in Fig. 30-62*b*). The α and β subunits are 26% identical in sequence except for a ~35-residue N-terminal segment of the α subunit, which the β subunit lacks. Eukaryotic 20S proteasomes are more complex in that they consist of multiple α-like and β-like subunits (7 of each type in yeast) that exhibit up to 5 different peptidase activities (which preferentially cleave after acidic, basic, and hydrophobic residues) vs only one of each type for the *T. acidophilum* 20S proteosome, which cleaves proteins rather nonspecifically.

The X-ray structure of the 20S proteasome from *T. acidophilum* reveals that its two inner rings each consist of 7 β subunits and its two outer rings each consist of 7 α subunits. Thus the overall structure of the 20S proteasome superficially resembles that of the unrelated molecular chaperone GroEL (Section 8-1C and Section 8-1a of this Supplement). The structures of the α and β subunits are remarkably similar except, of course, for the α subunit's N-terminal segment, which contacts the N-terminal segment in an adjacent α subunit. This accounts for the observation that α subunits alone spontaneously assemble into 7-membered rings (a capacity that is abolished by the deletion of of their N-terminal 35 residues), whereas β subunits alone remain monomeric.

The central channel of the 20S proteasome, which has a maximum diameter of 53 Å, consists of 3 large chambers, two of which are located at the interfaces between adjoining rings of α and β subunits, and the third larger chamber is located at the center of the protein between the two rings of β subunits. Access to these chambers is controlled by narrow constrictions at the centers of each of the four rings that are lined with hydrophobic residues. These allow only unfolded proteins to enter the central chamber in which, as we shall see below, the proteasome's active sites are located, thereby protecting properly folded proteins from indiscriminant degradation.

The X-ray structure of the 20S proteasome in complex with the aldehyde inhibitor acetyl-Leu-Leu-norleucinal (LLnL) revealed that the active sites are on the inner surface of the rings of β subunits, with the aldehyde function of the LLnL close to the side chain of Thr 1β. Deletion of this Thr or its mutation to Ala yields properly assembled 20S proteasomes that are completely inactive. This has led to the proposal that the 20S proteasome has a novel proteolytic mechanism in which the hydroxyl group of its Thr 1β is the attacking nucleophile.

Löwe, J., Stock, D., Jap, B., Zwicki, P., Baumeister, W., and Huber, R., Crystal structure of the 20S proteasome from the archaeon *T. acidophilum* at 3.4 Å resolution, *Science* **268**, 533–539 (1995).

Seemüller, E., Lupas, A., Stock, D., Löwe, J., Huber, R., and Baumeister, W., Proteasome from *Thermoplasma acidophilum*: A threonine protease, *Science* **268**, 579–582 (1995).

Jentsch, A. and Schlenker, S., Selective protein degradation: A journey's end within the proteasome, *Cell* **82**, 881–884 (1995).

(b) A Tagging System for the Degradation of Proteins Synthesized from Damaged mRNA

Ribosomes translating nuclease-damaged mRNAs that lack their 3' stop codons may stall when they reach the 3' end of this mRNA because, in the absence of stop codons, release factors cannot trigger the release of newly synthesized polypeptides from the ribosome. How does the cell prevent this from occurring? The answer to this question began to emerge with the observation that when mouse **interleukin-6 (IL-6;** a protein growth factor) was expressed at high levels in *E. coli,* the small population of polypeptides that were variably truncated at their C-termini all had the same C-terminal sequence, AANDENYALAA. This sequence is not encoded by IL-6 mRNA but, except for the first Ala, is encoded by a 362-nt RNA known as **10Sa RNA,** the product of the *ssrA* gene. The 3' end of the 10Sa RNA has a tRNA-like sequence and, in fact, can be charged with Ala. Moreover, the 10Sa RNA sequence that encodes the 10-residue peptide is immediately followed by a UAA stop codon. This suggests that a ribosome stalled at the 3' end of an mRNA that lacks a stop codon binds Ala-charged 10Sa RNA, appends this Ala to the end of its nascent polypeptide chain, switches to the 10Sa RNA to translate its encoded C-terminal peptide, and then normally terminates transcription at the 10Sa RNA's UAA stop codon.

The function of this novel protein-tagging system became clear when it was realized that the C-terminal residues of the 10Sa RNA-encoded peptide tag (YALAA) resemble the C-terminal sequence (WVAAA) recognized by the tail-specific periplasmic endoprotease **Tsp** as well as by a poorly characterized cytosplasmic protease that also degrades proteins in a tail-dependent manner. The system apparently functions to identify improperly translated proteins for degradation by Tsp and its cytoplasmic counterpart. Indeed, *ssrA⁻ E. coli* grow more slowly than do wild-type cells, presumably reflecting the presence of stalled ribosomes as well as the slow degradation of damaged and hence deleterious proteins.

To verify the above model, a transcriptional terminator was inserted in 3' regions of the genes encoding cytochrome b_{562} (a periplasmic protein) and the λ repressor (a cytoplasmic protein), upstream of their stop codons. The resulting polypeptides were indeed translated with the forgoing C-terminal tag. These proteins were degraded with half-lives of <0.5 min, whereas proteins constructed with the C-terminal control tag AANDENYYYALDD, which Tsp and its cytoplasmic counterpart do not recognize, had half-lives of >1 hr. Moreover, when an AANDENYALAA-tagged cytochrome b_{562} was expressed in *tsp⁻* cells, it also had a half-life of >1 hr.

Both Tsp and 10Sa RNA have homologs in gram-positive and gram-negative bacteria. It remains to be seen, however, whether this novel cotranslational tagging/degradation system is present in eukaryotes, although eukaryotes posttranslationally tag proteins with ubiquitin to mark them for degradation by the 26S proteasome.

Keiler, K.C., Waller, P.R.H., and Sauer, R.T., Role of a peptide tagging system in degradation of proteins synthesized from damaged messenger RNA, *Science* **271,** 990–993 (1996).

Tu, G.-F., Reid, G.E., Zhang, J.-G., Moritz, R.L., and Simpson, R.J., C-terminal extension of truncated recombinant proteins in *Eschericia coli* with a 10Sa decapeptide, *J. Biol. Chem.* **270,** 9322–9326 (1995).

(c) Addiional References

Hochstrasser, M., Protein degradation or regulation: Ub the judge, *Cell* **84,** 813–815 (1996). [Discuusses the regulatory functions of protein ubiqitination.]

Chapter 31

DNA REPLICATION, REPAIR, AND RECOMBINATION

2. ENZYMES OF REPLICATION

(a) The X-ray Structure of the Single-Strand Binding Protein Complexed to DNA

Single-strand binding protein (SSB) functions during DNA replication to prevent strand-separated DNA from reannealing. It also prevents the DNA from forming fortuitous intramolecular secondary structures, protects it from nucleases, and facilitates general recombination (Section 31-6A). The SSB encoded by bacteriophage T4, which is named **gp32,** consists of 301 residues. Proteolytic removal of its N-terminal 17 residues abolishes gp32:gp32 cooperativity (recall that single-stranded DNA becomes coated with multiple SSB subunits), whereas its C-terminal 46 residues are essential to the interaction of gp32 with other proteins such as T4 DNA polymerase. Removal of both of these terminal segments yields ''core gp32'', which binds to single-stranded DNA with the same affinity as does intact gp32.

The X-ray structures of core gp32 (residues 21-254) in its complexes with the single-stranded hexadeoxynucleotides pTTTTTT, pTTATTT, and pTTTATT reveal that the protein consists of three subdomains: A Zn^{2+}-binding subdomain, in which the Zn^{2+} is tetrahedrally liganded by three Cys side chains and a His side chain (removal of the Zn^{2+} greatly reduces the protein's thermostability and makes it hypersensitive to proteases, suggesting that the Zn^{2+} functions to stabilize the structure of gp32); a subdomain that consists mainly of a 5-stranded β sheet; and a 3-stranded subdomain that links the other two. The three subdomains are arranged to form a deeply clefted protein in which much of the cleft is lined with positively charged side chains. The hexanucleotides are largely disordered but, collectively, reveal that the DNA binds in the cleft against its positively-charged surface. The central portion is this cleft narrows to 15 Å, which explains why gp32 binds to single-stranded DNA with 10,000-fold greater affinity than to double-stranded DNA (B-DNA has a diameter of ~20 Å). The disorder of the bound hexanucleotides may reflect gp32's function of binding to single stranded DNAs of all sequences and freely sliding along them.

Shamoo, Y., Friedman, A.M., Parsons, M.R., Konigsberg, W.H., and Steitz, T.A., Crystal structure of a replication fork single-stranded DNA binding protein (T4 gp32) complexed to DNA, *Nature* **376,** 362–366 (1995); *and* Erratum, *Nature* **376,** 616 (1995).

(b) Additional References

Kim, Y., Eom, S.H., Wang, J., Lee, D.S., Suh, S.W., and Steitz, T.A., Crystal structure of *Thermus aquaticus* DNA polymerase, *Nature* **376,** 612–616 (1995). [The X-ray structure of the thermostable *Taq* polymerase (Section 28-8D) consists of an N-terminal 5'→3'-exonuclease domain, the first whose structure has been determined; a C-terminal polymerase domain whose structure is nearly identical to that of Klenow fragment; and an intervening domain which resembles the 3'→5'-exonuclease domain of Klenow fragment but which is nonfunctional.].

Arnold, E., Ding, J., Hughes, S.H., and Hostomsky, Z., Structures of DNA and RNA polymerases and their interactions with nucleic acid substrates, *Curr. Opin. Struct. Biol.* **5,** 27–38 (1995).

Sousa, R., Structural and mechanistic relationships between nucleic acid polymerases, *Trends Biochem. Sci.* **21,** 186–190 (1996).

3. PROKARYOTIC REPLICATION MECHANISMS

(a) Loading and Unloading the β Clamp

The replication of *E. coli* DNA proceeds at the rate of ~1000 nt/s. Thus, in lagging strand synthesis, the processive DNA replication machinery must be loaded and unloaded from the template strand every 1 to 2 s (Okazaki fragments are 1000 to 2000 nt in length). How does this occur?

The DNA polymerase III (Pol III) holoenzyme consists of the Pol III core ($\alpha\epsilon\theta$), which contains the complex's DNA polymerase (α) and 3'→5'-exonuclease (ϵ) functions but, by itself, is poorly processive; two copies of the β subunit, which forms the so-called β **clamp,** a ring around the DNA (Fig. 31-4) that renders its associated Pol III core infinitely processive; and the γ complex ($\gamma_2\delta\delta'\chi\psi$), which functions to load the β clamp around DNA in an ATP-dependent manner as well as to unload it and hence is also known as the **clamp loader.** A protein footprinting technique developed for the purpose as well as site-specific mutagenesis studies have shown that the C-terminal segment of the β subunit interacts with both the δ subunit of the clamp loader and the α subunit of the Pol III core.

The competition between the clamp loader and the Pol III core for the β clamp is modulated by DNA. In the absence of DNA, the β clamp associates with the clamp loader. However, after the β clamp has been loaded onto the DNA, the Pol III core binds to the β clamp more tightly than does the clamp loader, thereby displacing it and permitting processive DNA synthesis to proceed. Upon encountering the beginning of the previously synthesized Okazaki fragment, the Pol III core drops off the DNA and loses affinity for its associated β clamp. This again gives the clamp loader access to the β clamp, which it unloads from the DNA, thus recycling it for future use. The clamp loader and the Pol III core can then rapidly initiate the synthesis of a new Okazaki fragment because they are held in the vicinity of the lagging strand template through their linkage to the Pol III core engaged in leading strand synthesis (Fig. 31-22).

Herendeen, D.R. and Kelly, T.J., DNA polymerase III: Running rings around the fork, *Cell* **84,**

4–8 (1996).

Kelman, Z. and O'Donnell, M., DNA polymerase III holoenzyme: Structure and function of a chromosomal replicating machine, *Annu. Rev. Biochem.* **64,** 171–200 (1995).

Naktinis, V., Turner, J., and O'Donnel, M., A molecular switch in a replication machine defined by an internal competition for protein rings, *Cell* **84,** 137–145 (1996).

(b) Regulation of Chromosomal DNA Replication through Sequestration

The replication of the *E. coli* chromosome is initiated precisely once per cell generation at its replication origin site, *ori*C. Even in cells that contain multiple *ori*C sites, DNA replication is simultaneously initiated at each such site once and only once for every cell generation. After replication initiation has occurred, chain elongation proceeds at a uniform, largely uncontrolled rate. Thus, the control of chromosomal DNA replication occurs at its initiation stage.

Once initiation has occurred, the *ori*C site is somehow sequestered (segregated) from the replication initiation machinery for the remainder of the cell cycle. What is the mechanism of this **sequestration?** The 245-bp *ori*C site contains numerous palindromic GATC segments, each of which is specifically methylated at the N6 position of both of its adenines by *E. coli*'s Dam methyltransferase (Section 31-7). Thus, upon replication, the once fully methylated (on both strands) GATC segments become hemimethylated, that is, the newly synthesized GATC sequences are not yet methylated. Thus, the observation that membranes can bind hemimethylated *ori*C *in vivo* and *in vitro,* but not fully methylated or unmethylated *ori*C, led to the idea that sequestration occurs because hemimethylated *ori*C is bound to the membrane, thereby making it inaccessible to both the initiation machinery and the Dam methyltransferase. The *ori*C sites are released from sequestration only after the cell has reached a state in which replication initiation no longer occurs. They are then fully methylated in preparation for the next round of DNA replication.

The *seq*A gene encodes a 181-residue protein known as **SeqA.** This protein has been implicated in sequestration through several criteria. In *seq*A⁻ cells: the time to remethylate the GATC sites in *ori*C is reduced from 13 to 5 min, whereas the time to remethylate GATC sites in other positions is unaffected; the synchrony of initiation of multiple *ori*C sites is lost; and in the absence of a functional Dam methylase, fully methylated *ori*C-containing plasmids are efficiently propagated, whereas in the presence of SeqA they are replicated only once. SeqA has been purified and shown, through gel electrophoretic mobility shift (gel retardation) assays, to have a strong sequence-specific affinity for fully methylated *ori*C, and an even stronger affinity for hemimethylated *ori*C although without sequence specificity (except for the presence of GATC sites). However, the association of hemimethylated *ori*C with membrane requires the presence of SeqA. The SeqA molecules that bind to fully methylated *ori*C do so mostly along its "left" half, where they appear to interfere with the formation of the open complex (Fig. 31-23), thereby inhibiting replication initiation.

*ori*C is released from sequestration after about one-third of a cell generation time so that sequestration cannot determine the timing of replication initiation. Hence, there must be additional mechanisms that prevent initiation at *ori*C sites for the remaining two-thirds of the cell cycle.

Lu, M., Campbell, J.L., Boye, E., and Kleckner, N., SeqA: A negative modulator of replication initiation in E. coli, *Cell* **77**, 413–426 (1994).

Slater, S., Wold, S., Lu, M., Boye, E., Skarstad, K., and Kleckner, N., E. coli SeqA protein binds *ori*C in two different methyl-modulated reactions appropriate to its roles in DNA replication initiation and origin sequestration, *Cell* **82**, 927–936 (1995).

Crooke, E., Regulation of chromosomal replication in E. coli: sequestration and beyond, *Cell* **82**, 877–880 (1995).

(c) Additional References

Baker, T.A., Replication arrest, *Cell* **80**, 521–524 (1995).

4. EUKARYOTIC REPLICATION MECHANISMS

(a) PCNA Has a Similar Ring-Like Structure to That of the β Subunit of *E. coli* Pol III Holoenzyme

Mammalian proliferating cell nuclear antigen (PCNA), so-named because it was initially characterized as a cell cycle-dependent antigen, functions as the sliding clamp for DNA polymerase δ. Although biochemical evidence clearly indicates that PCNA (a trimer of identical 258-residue subunits in yeast) has a similar function to that of the β subunit of the *E. coli* DNA polymerase III holoenzyme (a dimer of identical 366-residue subunits), these two proteins exhibit essentially no discernable sequence identity. Nevertheless, the X-ray structure of yeast PCNA reveals that the two proteins have remarkably similar structures.

The PCNA trimer forms a 3-fold symmetric toroid that is almost identical in appearance to the 2-fold symmetric toroid formed by the β subunit dimer (Fig. 31-14). Each PCNA subunit consists of two topologically identical domains, each of which is composed of two strucurally similar subdomains. Thus, the PCNA trimer, as does the β subunit dimer, has pseudo 12-fold symmetry.

Even with the unambiguous sequence alignments dictated by their X-ray structures, sequence identities between the domains of PCNA, the domains of the β subunit, and the domains of both proteins range from 6% to 15%. This is well below the level required for reliable sequence alignment in the absence of structural information.

The PCNA toroid has inner and outer diameters of ~34 and ~80 Å vs ~38 and ~80 Å for the β subunit and hence both proteins can easily enclose a ~20-Å in diameter B-DNA molecule. The PCNA trimer has 81 Asp, 57 Glu, 54 Lys, and 24 Arg residues and hence has a net negative charge of −60 vs −22 for the β subunit under physiological conditions. Nevertheless, as in the β subunit, PCNA's Lys and Arg residues are concentrated around its inner surface, which is therefore positively charged. As in the β subunit, this inner surface is formed by 12 helices oriented so as to minimize their interactions with the grooves of a B-DNA that is threaded through the hole. Apparently, the eukaryotic PCNA has quite similar functions to the prokaryotic β subunit.

Krishna, T.S.R., Kong, X.-P., Gary, S., Burgers, P.M., and Kuriyan, J., Crystal structure of the eukaryotic DNA polymerase processivity factor PCNA, *Cell* **79,** 1233–1243 (1994).

(b) The C-Rich Strands of Telomere DNA Can Form 4-Stranded Helices

The 3'-ending strand of telomere DNA has a species-specific G-rich repeating sequence resembling TTGGGG. X-ray and NMR analyses have revealed that DNA segments with such sequences form 4-stranded helices in which the G bases associate as cyclic tetramers, known as G quartets, whose adjacent G residues hydrogen bond with their neighbors in a Hoogsteen-like arrangement (Fig. 31-35). However, telomere DNA, except for a 12- to 16-nt overhang on its 3'-end, is accompanied by its complementary strand. What is the structure of this C-rich strand (which has a repeating sequence resembling 5'-CCCCAA-3')?

It has long been known that two cytosines can base pair when one C but not the other is protonated at its N3 atom and the bases are oppositely oriented. Then, the amino group substituent to each atom C4 donates a hydrogen bond to O2 on the opposite base, and the N3$^+$—H group donates a hydrogen bond to N3 on the opposite base to form a triply hydrogen bonded C·C$^+$ base pair.

The X-ray structures of the DNA segments CCCT, TAACCC, and CCCAAT reveal that the C-rich strand of telomere DNA can also form a 4-stranded base paired helix but one whose character is very different from than that of the G-rich strand. Each C in these CCC segments forms a C·C$^+$ base pair with a C in a parallel CCC segment. Two such duplexes, in antiparallel orientation, intercalate through each other, much like two interpenetrating ladders, thereby forming a 4-stranded helix in which the C·C$^+$ base pairs from one double strand alternate with those from the other to form a stack of 6 C·C$^+$ base pairs. A similar 4-stranded intercalated structure had been previously observed in the NMR structure of d(TCCCCC) under acidic conditions. Thus, we are left with the intriguing possibility that the complementary strands of telomere DNA form two different types of 4-stranded helices *in vivo*.

Kang, C.H., Berger, I., Lockshin, C., Ratliff, R., Moyzis, R., and Rich, A., Crystal structure of intercalated four-stranded d(C$_3$T) at 1.4 Å resolution, *Proc. Natl. Acad. Sci.* **91,** 11636–11640 (1994); *and* Stable loop in the crystal structure of the intercalated four-stranded cytosine-rich metazoan telomere, *Proc. Natl. Acad. Sci.* **92,** 3874–3878 (1995).

Berger, I., Kang, C.H., Fredian, A., Ratliff, R., Moyzis, R., and Rich, A., Extension of the four-stranded intercalated cytosine motif by adenine·adenine base pairing in the crystal structure of d(CCCAAT), *Nature Struct. Biol.* **2,** 416–425 (1995).

Gehring, K., Leroy, J.-L., and Guéron, M., A tetrameric DNA structure with protonated cytosine·cytosine base pairs, *Nature* **363,** 561–565 (1993). [An NMR structure.]

(c) Additional References

Georgiadis, M.M., Jessen, S.M., Ogata, C.M., Telesnitsky, A., Goff, S.P., and Hendrickson, W.A., Mechanistic implication from the structure of a catalytic fragment of Moloney murine leukemia virus reverse transcriptase, *Structure* **3,** 879–892 (1995). [The X-ray structure of a weakly catalytically active tryptic fragment (residues 10-278 of a 671-residue protein) of **Moloney murine leukemia virus (MMLV) reverse transcriptase,** which includes its fingers and palm domain.]

Greider, C.W. and Blackburn, E.H., Telomeres, telomerase, and cancer, *Sci. Am.* **274**(2): 92–97 (1995).

Harley, C.B. and Villeponteau, B., Telomeres and telomerase in aging and cancer, *Curr. Opin. Genet. Dev.* **5,** 249–255 (1995).

Laughlan, G., Murchie, A.I.H., Norman, D.G., Moore, M.H., Moody, P.C.E., Lilley, D.M.J., and Luisi, B., The high-resolution crystal structure of a parallel-stranded guanine tetraplex, *Science* **265,** 520–524 (1995). [The X-ray structure of d(TGGGT).]

Pelletier, H., Sawaya, M.R., Kumar, A., Wilson, S.H., and Kraut, J., Structures of ternary complexes of rat DNA polymerase β, a DNA template-primer, and ddCTP, *Science* **264,** 1891–1903 (1994).

Rhodes, D. and Giraldo, R., Telomere structure and function, *Curr. Opin. Struct. Biol.* **5,** 311–322 (1995). [A review.]

Smith, F.W., Schultze, P., and Feigon, J., Solution structures of unimolecular quadruplexes formed by oligonucleotides containing *Oxytricha* telomere repeats, *Structure* **3,** 997–1008 (1995).

Sugino, A., Yeast DNA polymerases and their role at the replication fork, *Trends Biochem. Sci.* **20,** 319–323 (1995).

Wang, Y. and Patel, D.J., Solution structure of the Tetrahymena telomeric repeat d(T$_2$G$_4$)$_4$ G-tetraplex, *Structure* **2,** 1141–1156 (1994). [Reveals the structure of a single-stranded polynucleotide forming three stacked intrastrand G-quartets.]

5. REPAIR OF DNA

(a) X-Ray Structure of *E. coli* DNA Photolyase

A prominent form of DNA damage is the UV-induced formation of intrastrand pyrimidine dimers such as the thymine dimer (Fig. 31-36). DNA photolyase functions to restore a pyrimidine dimer to its original monomeric state through the application of the energy of an absorbed photon of blue light (~400 nm). Photolyases, which occur in many prokaryotes and eukaryotes (including goldfish, rattlesnakes, and marsupials but not humans), each contain two prosthetic groups: a light-absorbing cofactor, which depending on the species, may be either N^5,N^{10}-methenyltetrahydrofolate (N^5,N^{10}-methenyl-THF; Fig. 24-39) or a 5-deazariboflavin derivative (*p.* 1046); and FADH$^-$, to which the excitation is transferred and which in turn relays it to the pyrimidine dimer in the form of an electron, transiently yielding FADH$^{\cdot}$ (although, in the N^5,N^{10}-methenyl-THF-containing enzyme, the FADH$^-$ directly absorbs 20% of the photons it transfers to the substrate and hence the enzyme is active, albeit at a reduced level, in the absence of bound N^5,N^{10}-methenyl-THF).

E. coli DNA photolyase is a 471-residue monomeric protein whose light-harvesting cofactor is N^5,N^{10}-methenyl-THF. The X-ray structure of this enzyme reveals that it is folded into two domains with the N^5,N^{10}-methenyl-THF bound between them. The FADH⁻ adopts a U-shaped conformation such that its isoalloxazine and adenine rings are in close proximity, a conformation that differs significantly from the largely extended conformations assumed by the FADs in all other FAD–protein complexes of known structure. The centers of mass of the THF and isoalloxazine rings are ~17 Å apart, which permits energy transfer between them via a radiationless resonant mechanism with the observed 62% efficiency and 200 ps time constant.

DNA photolyase binds either double-stranded or single-stranded DNAs with high affinity but without regard to base sequence. The DNA binding site appears to be a positively charged flat surface on the enzyme. This surface is penetrated by a hole that has a size and polarity complementary to that of a pyrimidine dimer-containing dinucleotide. A pyrimidine dimer bound in the hole would be in van der Waals contact with the isoalloxazine ring and hence properly situated for high efficiency electron transfer from this ring system. This implies that a pyrimidine dimer in a double helix flips out of the helix in a manner similar to that observed in the structure of the modification methylase M.HhaI in complex with its target DNA (Fig. 31-66). The flip-out is probably facilitated by the relatively weak base pairing interactions of the pyrimidine dimer and the distortions it imposes on the double helix.

Park, H.-W., Kim, S.-T., Sancar, A., and Deisenhofer, J., Crystal structure of DNA photolyase from *Eschericia coli, Science* **268,** 1866–1872 (1995).

(b) The X-ray Structure of the Pyrimidine Dimer Excision Repair Enzyme T4 Endonuclease V in Complex with a Duplex DNA Containing a Thymine Dimer

An alternative way in which a pyrimidine dimer may be repaired is to replace it with the corresponding normally-linked nucleotides The first steps of this process, the excision of the pyrimidine dimer from the DNA, is catalyzed, in bacteriophage T4-infected *E. coli*, by **T4 endonuclease V (endo V)**. This 138-residue monomeric protein, despite its small size, does so through two distinct catalytic activities: (1) a pyrimidine dimer-specific glycosylase that excises the pyrimidine base on the dimer's 5' side yielding an AP (apurinic or apyrimidinic) site; and (2) an AP endonuclease that cleaves the phosphodiester bond on the 3' side of the AP site in a reaction that opens up the AP ribose ring yielding an α,β-unsaturated aldehyde.

The X-ray structure of endo V in complex with a thymine dimer-containing 14-bp duplex DNA (the sequence of whose thymine dimer-containing strand is: ATCGCGT^TGCGCT) has been determined. The protein, whose structure in the complex is essentially identical to that seen in the X-ray structure of the uncomplexed protein, consists of a single compact domain with a comma-like shape that contains 3 noncontacting α helices but no β sheets. The protein's positively charged concave face binds to the minor groove side of the DNA.

The complex's most striking feature is that the otherwise largely undistorted B-DNA is kinked by ~60° towards its major groove side (away from the protein) at the 5'-T of the centrally located thymine dimer. The A base opposite this 5'-T has flipped out of the double helix into a closely-fitting pocket on the surface of the protein, much like the flipped out f⁵C base observed in the structure of the modification methylase M.HhaI in complex with its target DNA (Fig. 31-66). Curiously, endo V does not contact the thymine dimer itself although it extensively interacts with the distorted sugar-phosphate backbone segment linked to the thymine dimer (whose P–P distance, for example, is 1.5 Å shorter than the 3.4-Å distance characteristic of canonical B-DNA due to a similar shortening of the distance between the covalently linked T bases in the dimer relative to normally stacked bases). The 10 × 9-Å hole between the two strands of the DNA arising, in part, from the absence of the flipped-out A base is filled by several of the enzyme's catalytically implicated side chains. Since flipped-out bases have now been seen or implicated in the structures of three DNA repair enzymes (see below) and two DNA methylases (see Section 31-7a of this Supplement), such a DNA deformation may be a general feature of enzymes that modify DNA bases.

Vassylyev, D.G., Kashiwagi, T., Mikami, Y., Ariyoshi, M., Iwai, S., Ohtsuka, E., and Morikawa, K., Atomic model of a pyrimidine dimer excision repair enzyme complexed with a DNA substrate: structural basis for damaged DNA recognition, *Cell* **83,** 773–782 (1995).

Morikawa, K., Ariyoshi, M.,Vassylyev, D.G.,Matsumoto, O., Katayanagi, K., and Ohtsuka, E., Crystal structure of a pyrimidine dimer-specific excision repair enzyme from bacteriophage T4: Refinement at 1.45 Å and X-ray analysis of three active site mutants, *J. Mol. Biol.* **249,** 360–375 (1995).

(c) The Structural Basis for the Specificity of Uracil-*N*-Glycosylase

Uracil-*N*-glycosylases [alternatively known as **uracil–DNA glycosylases (UDGases)**] are ubiquitous and highly conserved DNA-repair enzymes. They function to hydrolytically excise the uracil bases from DNA, leaving an AP (apurinic or apyrimidinic) site, which is subsequently excised by an AP-endonuclease and an exonuclease. The resulting gap is filled in by DNA polymerase and DNA ligase. Uracil bases in DNA arise in two ways: (1) the deamination of cytosine bases, which spontaneously occurs at the rate of ~120 times per day in each human cell and if uncorrected, would lead to $C \rightarrow T$ mutations in the next round of DNA replication; and (2) the occasional misincorporation into DNA of dUMP rather than dTMP, which although U can adequately base pair with A in DNA, would substantially perturb the interactions of regulatory segments of DNA with sequence-specific proteins and hence impair gene expression.

Uracil and thymine differ only by the replacement of uracil's H5 atom by a methyl group in thymine. Hence, UDGases must be exquisitely specific for uracil; otherwise they would severely damage DNA. In fact, UDGases excise only uracil from single and double stranded DNAs, with a preference for single-stranded DNA, and are inactive against the uracils in RNA.

Several X-ray structures of UDGases have been determined: that from **herpes simplex virus type 1 [HSV-1; 244 residues** (This virus lies dormant, sometimes for years, in mature human neurons, which lack UDGase. Hence, HIV-1 UDGase is important for viral reactivation and therefore is a target for anti-herpes drugs.)], alone and in complex with uracil and with d(pTTT); and those of the 39% identical human UDGase (a 224-residue truncated protein containing only residues 82-304 of the native enzyme), alone and in complex with **6-aminouracil,** and with a **UDGase inhibitor** encoded by the *Bacillus subtilis* bacteriophage **PBS2** (whose DNA contains uracil rather than thymine and hence must be protected against host UDGase). Both UDGases are monomeric proteins that have a 4-stranded parallel β sheet core flanked on both sides by helices. A sinuous positively charged channel runs along the protein surface over the C-terminal ends of the β strands and has a rigid pocket near one end. The uracil in the HSV-1 complex and the 6-aminouracil in the human complex bind quite snugly in the pocket such that the uracil ring is stacked on a conserved Phe side chain, its ''Watson–Crick'' edge is elaborately hydrogen bonded to conserved side chains, and its 5-position (to which the methyl group in thymine would be linked) is tightly abutted against the face of a Tyr phenyl ring that appears to be rigidly held in place byextensive associations with other conserved groups in the protein. Thus, T, A, and G are sterically excluded from this pocket, whereas C could not properly hydrogen bond to it. Highly conserved residues that have been implicated in catalysis through mutagenesis studies are located near the periphery of the pocket within "striking" distance of the uracil's attached C1' (which was modeled into the complex).

In the complex of d(pTTT) with HIV-1 UDGase, the trinucleotide lies partially in the groove with its 3'-T base bound in the mouth of the uracil-binding pocket rather than occupying it. Nevertheless, this thymine base appears to interact strongly with UDGase with its N3 forming a hydrogen bond with an Asp side chain and with its 5-methyl group in van der Waals contact with the edge of a Phe ring. It has therefore been proposed that this binding site for T functions as a specific "trap" for thymine, thereby preventing it from achieving even a low level of catalytically productive binding with the enzyme.

The structures of the uracil-containing complexes strongly indicate that a uracil in double-stranded DNA must flip out of the double helix in order to bind in UDGase's specificity pocket. This presumably happens in a manner similar to the flip-out of the methylated C base in the complex of M.HhaI with its target duplex DNA (Fig. 31-66) and with the A base in a thymine dimer-containing DNA in its complex with T4 endonuclease V (Section 31-5b of this Supplement), and hence would not greatly disturb the conformation of the DNA except around the flipped out base.

The complex of UDGase with the UDGase inhibitor (**Ugi**) from bacteriophage PBS2 reveals that Ugi, an 84-residue monomer, contains a 5-stranded antiparallel β sheet flanked by two helices. The β strand along one edge of the sheet is inserted into the UDGase's DNA-binding groove without contacting the uracil specificity pocket. The Ugi binds to the UDGase through shape, charge, and hydrogen bonding complementarity and thus appears to mimic duplex DNA. Leu 272, which protrudes from the surface of UDGase, an unusual position for a hydrophobic side chain, is inserted into a pocket in Ugi. Model building based on the structure of this complex suggests that Leu 272 acts to replace the flipped out uracil residue in the complex of UDGase with uracil-containing duplex DNA, in much the same way as Gln 237 replaces the flipped out f⁵C base in the X-ray structure of M.HhaI with its target DNA (Fig. 31-66).

Mol, C.D., Arvai, A.S., Sanderson, R.J., Slupphaug, G., Kavli, B., Alseth, I., Krokan, H.E., and Tainer, J.A., Crystal structure and mutational analysis of human uracil-DNA glycosylase: structural basis for specificity and catalysis, *Cell* **80,** 869–878 (1995).

Mol, C.D., Arvai, A.S., Sanderson, R.J., Slupphaug, G., Kavli, B., Krokan, H.E., Mosbaugh, D.W., and Tainer, J.A., Crystal structure of human uracil-DNA glycosylase in complex with a protein inhibitor: protein mimicry of DNA, *Cell* **82,** 701–708 (1995).

Savva, R., McAuley-Hecht, K., Brown, T., and Pearl, L., The structural basis for specific base-excision repair by uracil-DNA glycosylase, *Nature* **373,** 487–493 (1995).

(d) Additional References

Dodson, M.L., Michaels, M.L., and Lloyd, R.S., Unified catalytic mechanism for DNA glycosylases, *J. Biol. Chem.* **269,** 32709–32712 (1994).

Fogh, R.H., Ottleben, G., Rüterjans, H., Schnarr, M., Boelens, R., and Kaptein, R., Solution structure of the LexA repressor DNA binding domain determined by ^{1}H NMR spectroscopy, EMBO J., 13, 3936–3944 (1994). [The 84-residue N-terminal DNA-binding domain of LexA contains a helix–turn–helix (HTH) motif with an extra residue in its recognition helix. LexA's DNA-binding domain is topologically identical to those of catabolite gene activator protein (CAP; Fig. 29-22) and histone 5 (Fig. 33-11).]

Mol, C.D., Kuo, C.-F., Thayer, M.M., Cunningham, R.P., and Tainer, J.A., Structure and function of the multifunctional DNA-repair enzyme exonuclease III, *Nature* **374,** 381–386 (1995). [The multifunctional enzyme **exonuclease III** (not to be confused with *E. coli* endonuclease III) is *E. coli*'s major AP DNA-repair endonuclease, a 3'-repair diesterase, a 3'→5' exonuclease, a 3'-phosphomonoesterase, and an RNase.]

Pegg, A.E., Dolan, M.E., and Moschel, R.C., Structure, function, and inhibition of O^6-alkylguanine alkyltransferase, *Prog. Nucleic Acid Res. Mol. Biol.* **51,** 167223 (1995).

Sancar, A., Excision repair in mammalian cells, *J. Biol. Chem.* **270,** 15915–15918 (1995)

Thayer, M.M., Ahern, H., Xing, D., Cunningham, R.P., and Tainer, J.A., Novel DNA binding motifs in the DNA repair enzyme endonuclease III crystal structure, *EMBO J.* **14,** 4108–4120 (1995). [A higher resolution study of the structure shown in Fig. 31-40. Consideration of this structure and its catalytically implicated residues suggests that it binds DNA with a flipped out base bound within its active site pocket.]

Trends Biochem. Sci. **20**(10): (1995). [The October, 1995 issue contains a series of reviews of DNA repair.]

6. RECOMBINATION AND MOBILE GENETIC ELEMENTS

(a) Structure of γδ Resolvase in Complex with a 34-bp DNA Containing Its Cleavage Site

The γδ resolvase, which is encoded by the γδ transposon (a member of the Tn3 family of replicative transposons), catalyzes a site-specific recombination event in which a circular DNA containing two copies of the γδ transposon (a cointegrate) is "resolved", via double-strand DNA cleavage, strand exchange, and religation (the last step in Fig. 31-60), into two concatenated DNA circles that each containing one copy of the γδ transposon. Each γδ transposon contains a 114-bp *res* site that includes 3 binding sites for γδ resolvase dimers, each of which contains an inverted repeat of the γδ resolvase's 12-bp recognition sequence. γδ Resolvase is a dimer of identical 183-residue subunits, the structure of whose 140-residue N-terminal chymotryptic fragment had been previously determined (center of Fig. 31-64). This fragment contains the protein's dimerization interface, its active Ser 10 residue (which forms a transient phosphoSer bond with the 5' phosphate at the cleavage site), and residues that are important for higher order interactions between γδ resolvase dimers. The C-terminal chymotryptic fragment (residues 141-183), which contains sequences homologous to the helix–turn–helix (HTH) DNA-binding motif, specifically binds to the 12-bp recognition site.

The X-ray structure has been determined of γδ resolvase in complex with a 34-bp DNA segment containing an inverted repeat of the 12-bp recognition sequence separated by an 8-bp spacer. Each resolvase monomer consists of a catalytic domain (residues 1-120) whose structure closely resembles that observed in the absence of DNA, a C-terminal DNA-binding domain (residues 148-183) that forms a 3-helix bundle whose C-terminal two helices and their connecting segment comprise an HTH motif, and an extended arm (residues 121-147) that connects the N- and C-terminal domains and which is disordered in the absence of DNA.

The structure of the DNA complex resembles the model drawn in Fig. 31-64 (in which the γδ resolvase's N-terminal domain dimer is shown at the center): The N-terminal domain dimer approaches the DNA from its minor groove side along its local 2-fold axis; the C-terminal domains (which resemble the Hin domains shown in the top and bottom of Fig. 31-64) each bind in the major groove of their target sequence on the opposite side of the DNA from the N-terminal domain dimer such that they are separated by 2 helical turns; and the extended arms which connect the two domains each more or less follows the path indicated by the dashed lines in Fig. 31-64. There are, however, two major differences with the model shown in Fig. 31-64: (1) The DNA, which otherwise closely assumes the cannonical B-DNA conformation, is centrally kinked by ~60° such that it bends towards its major groove, away from the N-terminal domain dimer; and (2) residues 121 to 136, which were disordered in the structure of the N-terminal domain alone, continue its C-terminal helix (helix E) such that it binds over the minor groove of the DNA. The dimer's two E helices thereby grip the DNA like a pair of chopsticks in a manner reminiscent of the way the bZIP region of GCN4 interacts with DNA (Fig. 33-57; although bZIP GCN4 binds in DNA's major groove and its helices are more nearly parallel than are the E helices of γδ resolvase). The structure is asymmetric with Ser 10 on one monomer much closer to the DNA than the other although both are quite distant from the scissile bonds on the DNA. This suggests that the two single-strand cleavage reactions catalyzed by the dimer may occur sequentially and, in any case, require significant conformational changes.

The resolution of the cointegrate involves three γδ resolvase dimers bound to each of the cointegrate's *res* sites. Thus, although an understanding of the mechanism of this resolution reaction will require the knowledge of how these various γδ resolvase dimers interact, the present structure has provided insights on the structural requirements for this reaction.

Yang, W. and Steitz, T.A., Crystal structure of the site-specific recombinase γδ resolvase complexed with a 34 bp cleavage site, *Cell* **82,** 193–207 (1995).

(b) Additional References

Grindley, N.G.F. and Leschziner, A.E., DNA transposition: from a black box to a color monitor, *Cell* **83,** 1063–1066 (1995).

Rao, B.J., Chiu, S., Bazemore, L.R., Reddy, G., and Radding, C., How specific is the first step of homologous recombination, *Trends Biochem. Sci.* **20,** 109–113 (1995).

Rice, P., Craigie, R, and Davies, D.R., Retroviral integrases and their cousins, *Curr. Opin. Struct. Biol.* **6,** 76–83 (1996).

Sinagawa, H. and Iwasaki, H., Processing the Holliday junction in homologous recombination, *Trends Biochem. Sci.* **21,** 107–111 (1996).

7. DNA METHYLATION AND TRINUCLEOTIDE REPEAT EXPANSIONS

(a) The DNA Complexed to HaeIII Methyltransferase Has a Flipped-Out Cytosine Base

HaeIII methyltransferase (M.HaeIII) catalyzes the transfer of a methyl group from SAM to the 5-position of the underlined cytosine base of its palindromic 4-bp recognition site, GG<u>C</u>C. The X-ray structure was determined of this monomeric 330-residue protein in complex with an 18-bp DNA containing the protein's recognition sequence with the target C base replaced by the mechanism-based inhibitor 5-fluoro-C and the equivalent C on the opposite strand replaced by 5-methyl-C. The protein consists of a large catalytic domain (residues 1-182 and 306-275), which is closely superimposable on that of M.HhaI (Fig. 31-66), and a small recognition domain (residues 276-305), which exhibits little secondary structure and has only limited structural resemblance to the small domain of M.HhaI.

The DNA binds in the cleft formed by the two domains and their bridging segment (residues 183-275). As is seen in the structure of M.HhaI complexed to its target DNA, the C base to be methylated in the M.HaeIII complex is flipped out of the double helix and into a pocket in the protein that contains the catalytic groups. Since all of M.HaeIII's catalytic groups as well as the residues lining its SAM-binding pocket (which is empty in this structure) have conserved counterparts in the structure of M.HhaI, it is quite likely that the two proteins follow similar catalytic mechanisms.

The flipping out of the substrate C from the DNA is accompanied by an extensive rearrangement of its nearby bases. The G base that would otherwise be Watson–Crick paired to the substrate C shifts along the helix to form a highly propeller twisted (by 58°) Watson–Crick base pair with the C adjacent to the substrate C (the outer C in the GG<u>C</u>C recognition sequence). Consequently, the G that would normally be base paired to this outer C is no longer base paired. However, this G appears to be stabilized in its position through hydrogen bonding interactions with Arg 243. The shift of the out-of-register base-paired G unstacks it and opens an ~8-Å-wide cleft into which the side chain of Ile 221 protrudes, but not deeply so that a large solvent channel also runs through the DNA. This rearrangement would not be possible in the M.HhaI structure because its GC<u>G</u>C target sequence could not support the forgoing exchange of base pairing partners (i.e., it would have to form a G·G pair). Rather, the flipped-out C is replaced by a Gln side chain without significant distortion of the DNA (Fig. 31-66).

Reinisch, K.M., Chen, L. Verdine, G.L., and Lipscomb, W.N., Jr., The crystal structure of HaeIII methyltransferase complexed to DNA: An extrahelical cytosine and rearranged bases, *Cell* **82,** 143–153 (1995).

Roberts, R.J., On base flipping, *Cell* **82,** 9–12 (1995)

(b) Additional References

Cheng, X., Structure and function of DNA methyltransferases, *Annu. Rev. Biophys. Biomol. Struct.* **24,** 293–318 (1995).

Cross, S.H. and Bird, A.P. CpG islands, *Curr. Opin. Genet. Dev.* **5,** 309–314 (1995).

Oostra, B.A. and Willems, P.J., A fragile gene, *BioEssays* **17,** 941–947 (1995). [Discusses fragile

X syndrome.]

Warren, S.T., The expanding world of trinucleotide repeats, *Science* **271,** 1374–1375 (1996).

Wells, R.D., Molecular basis of genetic instability of triplet repeats, *J. Biol. Chem.* **271,** 2875–2878 (1996).

Chapter 32

VIRUSES: PARADIGMS FOR CELLULAR FUNCTION

2. SPHERICAL VIRUSES

(a) Structure of the MS2 Coat Protein Dimer in Complex with the Operator RNA of the Viral Replicase

The coat protein of the RNA bacteriophage MS2 (Fig. 32-33) serves double duty: A dimer of the 129-residue coat protein subunit binds to a 19-nt stem-and-loop structure containing the initiation codon of the viral replicase gene. This shuts off replicase synthesis, thereby stopping viral replication and facilitating virion assembly. In fact, the presence of this operator segment alone can trigger the assembly of the viral capsid at low protein concentrations.

The X-ray structure of the MS2 coat protein dimer in complex with this RNA segment represents the first known structure of a sequence-specific protein–RNA complex that is not an aminoacyl-tRNA synthetase·tRNA complex. The RNA consists of a 7-bp stem that contains a one-base loop-out and has a 4-nt loop. The three base pairs formed by the 5' and 3' ends of the stem are not visible in the X-ray structure. The remaining 13 nts form a crescent-shaped structure that binds across the bottom of the 8-stranded antiparallel β sheet formed by the MS2 coat protein dimer, whose structure is substantially the same as that in the intact virus. Only a few conserved residues that directly interact with the RNA bases essential for binding have slightly different conformations from those in the native virion.

Valegård, K., Murray, J.B., Stockley, P.G., Stonehouse, N.J., and Liljas, L., Crystal structure of an RNA bacteriophage coat protein–operator complex, *Nature* **371,** 623–626 (1994).

(b) Additional References

Ban, N. and McPherson, A., The structure of satellite panicum mosaic virus at 1.9 Å resolution, *Nature Struct. Biol.* **2,** 882–883 (1995). [The structure of the smallest virus yet determined, which has an 850-nt genome and a $T = 1$ capsid of 60 identical 157-residue subuits.]

Bhuvaneshwari, A., Subramnaya, H.S., Gopinath, K., Savrithi, H.S., Nayudu, M.V., and Murthy,

M.R.N., Structure of *sesbania* mosaic virus at 3 Å resolution, *Structure* **3,** 1021-1030 (1995). [The structure of a virus whose quaternary structure is similar to that of southern bean mosaic virus (SBMV).]

Chapman, M.S. and Rossmann, M.G., Single-stranded DNA-protein interactions in canine parvovirus, *Structure* **3,** 151–162 (1995).

Johnson, J., Functional implications of protein–protein interactions in icosahedral viruses, *Proc. Nat'l. Acad. Sci.* **93,** 27–33 (1996). [A review.]

Liljas, L., Viruses, *Curr. Opin. Struct. Biol.* **6,** 151–156 (1996); *and* Harrison, S.C., Virus structures and conformational rearrangements, *Curr. Opin. Struct. Biol.* **5,** 157–164 (1995).

Muckelbauer, J.K., Kremer, M., Minor, I., Diana, G., Dutko, F.J., Groarke, J., Pevear, D.C., and Rossmann, M.G., The structure of coxsackievirus B3 at 3.5 Å resolution, *Structure* **3,** 653–667 (1995). [A picornavirus.]

Speir, J.A., Munshi, S., Wang, G., Baker, T.S., and Johnson, J.E., Structures of the native and swollen forms of cowpea chlorotic mottle virus determined by X-ray crystallography and cryo-electron microscopy, *Structure* **3,** 63–78 (1995). [A $T = 3$ spherical virus.]

Stehle, T., Gamblin, S.J., Yan, Y., , and Harrison, S.C., The structure of simian virus 40 refined at 3.1 Å resolution, *Structure* **4,** 165–182 (1996).

Stehle, T. and Harrison, S.C., Crystal structure of murine polyomavirus in complex with straight-chain and branched-chain sialyloligosaccharide receptor fragments, *Structure* **4,** 183–194 (1996). [Murine polyomavirus is a homolog of SV40.]

Timmons, P.A.. Wild, D., and Witz, J., The three-dimensional distribution of RNA and protein in the interior of tomato bushy stunt virus: a neutron low-resolution single-crystal diffraction study, Structure 2, 1191–1201 (1994). [Confirms that the protein in TBSV is distributed in two shells with the RNA sandwiched between them (Fig. 32-17), but reveals that this distribution in the inner shells is far more inhomogeneous than was previously suspected.]

4. INFLUENZA VIRUS

Coleman, P.M., Influenza virus neuraminidase: Structure, antibodies, and inhibitors, *Protein Sci.* **3,** 1687–1696 (1994). [A review.]

5. SUBVIRAL PATHOGENS

(a) Aged Mice that Lack the PrP Gene Lose Certain Brain Neurons

Prion protein (PrP) is a glycoprotein that is expressed on neuronal cell surfaces, whose conformational conversion to a protease-resistant form has been implicated in the pathogenesis of transmissable spongiform encephalopathies such as scrapie and Creutzfeldt-Jacob disease (CJD). Thus it came as a surprise that mice which are homozygous for a disrupted form of the gene encoding PrP appear entirely normal and have apparently normal offspring. However, further investigations revealed that after a particular strain of PrP-null mice reached ~70 weeks of age, they exhibited progressive ataxia (loss of coordination of voluntary muscles). Pathological examination of these mice revealed a significant loss of cerebellar Purkinje cells (a type of neuron). Apparently, PrP plays a role in the long term survival of Purkinje neurons. However, two other strains of PrP-null mice had not developed ataxia at up to 93 weeks of age.

Sakaguchi, S., et al., Loss of cerebellar Purkinje cells in aged mice homozygous for a disrupted *PrP* gene, *Nature* **380,** 528–530 (1996).

(b) Host PrP^C Must Be Present for the Development of the Symptoms of Scrapie

Grafting neuronal tissue overexpressing PrP^C onto the brains of PrP-deficient mice followed by intracerebral innoculation of PrP^{Sc} led to the accumulation of high levels of PrP^{Sc} in the grafts, some of which migrated into the host brain. Nevertheless, these mice did not develop the symptoms of scrapie, even after 16 months. This indicates that PrP^{Sc} is not, by itself, neurotoxic when present outside neurons. Evidently, it is not the deposition of PrP^{Sc} that causes the symptoms of scrapie but rather the interaction of PrP^{Sc} with normally occurring PrP^C.

Brandner, S., Isenmann, S., Raeber, A., Fischer, M., Sailer, A., Kobayashi, Y., Marino, S., Weissmann, C., and Aguzzi, A., Normal host prion protein necessary for scrapie-induced neurotoxicity, *Nature* **379,** 339–343 (1996).

(c) Fatal Familial Insomnia Is Transmissable

Fatal familial insomnia (FFI) is an inherited human spongiform encephalopathy arising from a mutant form of the *Prn-p* gene. The intracerebral innoculation into mice of homogenized thalamus tissue from a victim of FFI resulted in symptoms typical of spongiform encephalopathies and death between 397 and 506 days after innoculation. The pathological examination of the brains of these mice revealed characteristics typical of spongiform encephalopathies. This establishes that FFI is a subtype of the transmissible spongiform encephalopathies.

Tateishi, J., Brown, P., Kitamoto, T., Hoque, Z.M., Roos, R., Wollman, R., Cervenáková, L., and Gajdusek, D.C., First experimental transmission of fatal familial insomnia, *Nature* **376,** 434–435 (1995).

(d) Additional References

Collinge, J., Palmer, M.S., Sidle, K.C.L., Hill, A.F., Gowland, I., Meads, J., Asante, E., Bradley, R., Doey, L.J., and Lantos, P.L., Unaltered susceptibility to BSE in transgenic mice expressing human prion protein, *Nature* **378,** 779–783 (1995).

Nguyen, J., Baldwin, M.A., Cohen, F.E., and Prusiner, S.B., Prion protein peptides induce α-helix to β-sheet conformational transitions, *Biochemistry* **34,** 4186–4192 (1995).

Telling, G.C., Scott, M., Mastrianni, J., Gabizon, R., Torchia, M., Cohen, F.E., DeArmond, S.J., and Pruisiner, S.B., Prion propagation in mice expressing human and chimeric PrP transgenes implicates the interaction of cellular PrP with another protein, *Cell* **83,** 79–90 (1995).

Chapter 33

EUKARYOTIC GENE EXPRESSION

1. CHROMOSOME STRUCTURE

Gruss, C. and Knippers, R., Structure of replicating chromatin, *Prog. Nucleic Acid Res. Mol. Biol.* **52,** 337–365 (1996).

van Holde, K. and Zlatanova, J., Chromatin higher order structure: chasing a mirage, *J. Biol. Chem.* **270,** 8373–8376 (1995).

Wolffe, A.P. and Pruss, D., Targeting chromatin disruption: Transcription regulators that acetylate histones, *Cell* **84,** 817–819 (1996).

Zlatanova, J. and van Holde, K., The linker histones and chromatin structure: new twists, *Prog. Nucleic Acid Res. Mol. Biol.* **52,** 217–259 (1996).

2. GENOMIC ORGANIZATION

(a) EKLF Is an Essential Transcription Factor for β-Globin Gene Production in Mice

Globin synthesis is regulated in a tissue-specific and a developmental stage-specific manner (Fig. 33-33). β-Globin is a component of adult hemoglobin and hence the gene encoding it is the last in the β-gene cluster to be activated. The upstream segments of the β-globin gene contain CACCC sequences that bind multiple copies of certain transcription factors, including the ubiquitously expressed Sp1 (Section 33-3B) and **erythroid Krüppel-like factor [EKLF** (*Krüppel* is a gene involved in differentiation; Section 33-4B)]. Embryonic mice made deficient in EKLF die during fetal liver erythropoiesis (Fig. 33-33, *top*) and exhibit the symptoms of severe β-globin deficiency seen in β-thalassemia. This suggests that EKLF may facilitate the fetal-to-adult (γ to β) hemoglobin switch in humans and therefore that selective inhibition of EKLF may help to maximize γ-globin expression in adults with thalasssemia or other hemoglobinopathies.

Perkins, A.C., Sharpe, A.H., and Orkin, S.H., Lethal β-thalaseammia in mice lacking the CACCC-transcription factor EKLF, *Nature* **375,** 319–322 (1995).

(b) Additional References

Orkin, S.H., Transcription factors and hematopoietic development, *J. Biol. Chem.* **270,** 4955–4958 (1995).

Struhl, K., Chromatin structure and RNA polymerase II connection: implications for transcription, *Cell* **84,** 179–182 (1996).

3. CONTROL OF EXPRESSION

(a) Nuclear Receptors

The nuclear receptor superfamily, which occurs in animals ranging from worms to humans, is composed of >150 proteins that bind a variety of hormones such as steroids (glucocorticoids, mineralocorticoids, progesterone, estrogens, and androgens), vitamin D_3, thyroid hormone (Section 34-4A), retinoids (Section 33-4B), and prostanoids (Section 23-7). These small lipophilic molecules readily pass through the plasma membrane into the cytoplasm where they bind to their corresponding nuclear receptors (which, in contrast to other hormone receptors, are soluble rather than membrane-bound). The resulting receptor-ligand complex migrates to the nucleus where it specifically binds to its target DNA sequence(s), known as a **hormone response element (HRE),** and transcriptionally activates the associated gene(s). The above hormones are thereby potent regulators of development, cell differentiation, and organ physiology.

The nuclear receptors have a common modular structure that consists of six functionally separable segments (designated A-F): a variable N-terminal region (A/B); a conserved DNA-binding domain (**DBD;** C), which binds to the corresponding HRE; a variable hinge region (D); a conserved ligand-binding domain (**LBD;** E), which also contains homo- and/or heterodimerization surfaces (see below) and a ligand-dependent transactivation function; and a variable C-terminal region (F). The DBD consists of two zinc fingers that have been structurally characterized in the steroid receptors (Fig. 33-54). The HREs to which they bind have the half-site (core) consensus sequences AGAACA for certain steroid receptors and AGGTCA for other nuclear receptors. These sequences are arranged in direct repeats ($\rightarrow n \rightarrow$), inverted repeats ($\rightarrow n \leftarrow$), or everted repeats ($\leftarrow n \rightarrow$), where n represents a 0- to 8-bp spacer (most commonly 1-5 bp) to whose length a specific receptor is targeted.

The nuclear receptors have been grouped into four classes according to their ligand-binding, DNA-binding, and dimerization characteristics: Class I receptors, which include the steroid hormone receptors, are ligand-induced homodimers that bind to DNA half-sites arranged in inverted repeats; Class II receptors, which include the receptors for thyroid hormone, all-trans retinoic acid (**T-RA,** p. 1180), vitamin D_3 (**VDR),** the prostanoids, and the insect steroid hormone ecdysone (p. 1159), heterodimerize with the so-called **retinoid X receptor (RXR)** (whose ligand is **9-cis retinoic acid)** and bind mostly to direct repeats; Class III receptors bind mostly to direct repeats as homodimers; and Class IV receptors bind to single core sites as monomers. The ligands of many of the Class III and Class IV nuclear receptors are unknown (these proteins were identified as nuclear receptors through sequence homology studies) and hence they are called **orphan receptors,** although it seems likely that many of them do not bind ligands and hence, for them, the term orphan receptor is a misnomer. The DBD of the Class IV receptors consists of the consensus core site AGGTCA that has a 5' extension of 2 to 4 nt.

The X-ray structures of three LBDs have been determined, those of human RXR-α, human **retinoic acid receptor (RAR)-γ** in complex with T-RA, and the rat α_1 **thyroid hormone receptor (TR)** in complex with the thyroid hormone agonist **3,5-dimethyl-3'-isopropylthyronine (Dimit,** an analog of the thyroid hormone T_3; *p.* 1264). These homologous ~250-residue domains have quite similar structures in which 10 helices are organized in a novel three-layered fold that is probably characteristic of LBD's. Although the intact TR–T_3 complex both homodimerizes and heterodimerizes with RXR on the appropriate HREs, the TR LBD–Dimit complex is a monomer in its crystal structure. Likewise, the RAR LBD crystallizes as a monomer as is in solution. However, the RXR LBD–T-RA forms a homodimer both in the crystal and in solution.

The RAR LBD binds its T-RA ligand in a completely buried hydrophobic pocket that closely corresponds to the putative binding pocket in the RXR LBD for its 9-cis-retinoic acid ligand. In fact, the major difference between the structures of the RAR and RXR LBDs is that the C-terminal helix, which extends away from the body of the RXR LBD, folds back towards the core of the RAR LBD so as to "seal" its T-RA ligand in its binding pocket. The observation that the opening sealed by the C-terminal helix appears to be the only way that T-RA can enter its binding pocket suggests that this conformational change is induced by ligand binding. Furthermore, since the C-terminal helix has been implicated in hormone-dependent transcriptional activation, it seems quite likely that this putative conformational change is responsible for converting the LBD into its transcriptionally active state. This model is supported by the structure of the TR LBD–Dimit complex, whose C-terminal helix similarly "seals" the Dimit into its fully buried hydrophobic binding pocket.

The X-ray structure of the complex of the DBDs of human RXR-α and human TR-β with a 16-bp DNA containing two direct AGGTCA repeats separated by 4 bp ($n = 4$) reveals how these DBDs interact with this **thyroid hormone response element (TRE).** The RXR and TR DBDs, which cooperatively dimerize on the surface of the TRE but not in solution, respectively bind to the TRE's 5' and 3' half-sites. The structures of the RXR and TR core DBDs and their disposition with respect to the DNA closely resemble those of the glucocorticoid receptor (GR; Fig. 33-54) and the estrogen receptor (ER) DBDs; all consist of two closely associated zinc modules in each of which a Zn^{2+} ion is coordinated by four Cys residues, and all interact with their respective HREs via a recognition helix from each subunit in successive major grooves on the same face of the DNA. However, the TR DBD contains a C-terminal extension not present in these other DBD-containing structures that, in part, forms a helix, its so-called A-helix, which greatly increases the TR–DNA interface by interacting, through direct and water-mediated hydrogen bonds, with the phosphates and base pair edges of the DNA's minor groove. This accounts for the observation that the TR DBD, unlike other nuclear receptor DBDs, can bind to DNA as a monomer and recognizes the DNA sequences on the 5' side of its consensus half-site.

Since the TRE contains direct repeats, the TRE and RXR DBDs interact in a head-to-tail arrangement rather than in the head-to-head arrangement observed in the structure of the GR DBD in its complex with the inverted repeat-containing GRE (Fig. 33-54*b*). Thus, the second zinc module of the RXR DBD associates with the first zinc module of the TR DBD over the minor groove of the TRE's 4-bp spacer.

RXR also forms heterodimers with RAR and VDR on HREs that differ from the TRE only in that the spacer connecting the two directly repeated AGGTCA half-sites consists of 3 and 5 bp, respectively, rather than 4 bp. Model building based on the close homology of the RAR and VDR DBDs with that of TR and employing energy minimization calculations suggest that the

RAR and VDR DBDs, when bound to their respective HREs, make favorable contacts with the RXR DBD. Model building also shows that the A-helix of the TR DBD would sterically clash with portions of the RXR DBD on HREs with spacers of 1, 2, or 3 bps. This explains, at least in part, how the various nuclear receptor heterodimers can discriminate among HREs that differ only in the length of the spacer between their directly repeated AGGTCA half-sites.

Beato, M., Herrlich, P., and Schütz, G., Steroid hormone receptors: many actors in search of a plot, *Cell* **83,** 851–857 (1995).

Bourguet, W., Ruff, M., Chambon, P., Gronemeyer, H., and Moras, D., Crystal structure of the ligand-binding domain of the human nuclear receptor RXR-α, *Nature* **375,** 377–382 (1995).

Rastinejad, F., Perlmann, T., Evans, R.M., and Sigler, P.B., Structural determinants of nuclear receptor assembly on DNA direct repeats, *Nature* **375,** 203–211 (1995). [The X-ray structure of the heterodimer of the RXR and TR DBDs in complex with the TRE.]

Gewirth, D.T. and Sigler, P.B., The basis for half-site specificity explored through a non-cognate steroid receptor-DNA complex, *Nature Struct. Biol.* **2,** 386–394 (1995). [The X-ray structure of a GR DBD, which has been mutagenized in three residues to give it ER-like sequence specificity, in complex with an ERE (which has an AGGTCA half-site vs AGAACA for the GRE) reveals a slight difference in the helical geometries of the ERE and GRE half sites. This prevents the formation of a side-chain–base contact, resulting in a gap that is filled by at least five water molecules, thereby imposing a potential entropic burden on the stability of the interface, which accounts for the lesser stability of the noncognate vs the cognate (GR–GRE and ER–ERE) interfaces.]

Kastner, P., Mark, M., and Chambon, P., Nonsteroid nuclear receptors: What are genetic studies telling us about their role in real life, *Cell* **83,** 859–869 (1995).

Mangelsdorf, D.J., et al., The nuclear receptor superfamily: the second decade, *Cell* **83,** 835–839 (1995).

Mangelsdorf, D.J., and Evans, R.M., The RXR heterodimers and orphan receptors, *Cell* **83,** 841–850 (1995).

Renaud, J.-P., Rochel, N., Ruff, M., Vivat, V., Chambon, P., Gronemeyer, H., and Moras, D., Crystal structure of the RAR-γ ligand-binding domain bound to all-*trans* retinoic acid, *Nature* **378,** 681–689 (1995).

Schwabe, J.W.R., Chapman, L., and Rhodes, D., The oestrogen receptor recognizes an imperfectly palindromic response element through an alternative side-chain conformation, *Structure* **3,** 201–213 (1995). [The X-ray structure of the ER DBD in complex with an ERE that has a single base substitution in one of its half sites (a change that decreases the affinity of the ER–ERE complex by only 10-fold) reveals that the recognition of the nonconsensus sequence is achieved by the rearrangement of a Lys side chain to make an alternative base contact. This suggests how transcription factors recognize and promote transcription from DNA targets that differ from their consensus sequences, as they do in most cases.]

Chapter 33. Eukaryotic Gene Expression

Wagner, R.L., Apriletti, J.W., McGrath, M.E., West, B.L., Baxter, J.D., and Fletterick, R.J., A structural role for hormone in the thyroid hormone receptor, *Nature* **378**, 690–697 (1995). [The X-ray structure of the TR LBD in complex with Dimit.]

(b) X-Ray Structure of NF-κB p50 in Complex with DNA

Nuclear factor κB (NF-κB), a protein that was originally identified in B-lymphocytes as an inducible nuclear activity which binds to the κB motif in the immunoglobulin κ light chain gene enhancer (B-cells and the immunoglobulin genes are discussed in Section 34-2A-C), is present in nearly all animal cells although its role is particularly prominent in the immune system. It is a heterodimer of the **p50** and **p65** (alternatively **RelA**) proteins, both of which contain a ~300-residue segment known as the **Rel homology region (RHR)** because it also occurs in the product of the *rel* oncogene. RHRs, which mediate protein dimerization, DNA binding, and contain nuclear localization signals (**NLSs;** Section 11-4C), are present in a variety of proteins that serve as regulators of cellular defense mechanisms against stress, injury, and external pathogens, as well as of differentiation. There are two classes of **Rel proteins:** those such as p50 and the closely related **p52,** which are generated by the proteolytic processing of a larger precursor; and those such as p65, **c-Rel,** and the *Drosophila* morphogen proteins **Dorsal** and **Dif** (proteins that mediate development; see Section 33-4B), which are not generated by proteolytic processing, whose N-terminal domains contain an RHR, and whose highly variable C-terminal domains are strong transcriptional activators.

The activity of NF-κB is primarily regulated by proteins known as **inhibitor-κBs (IκBs),** which by binding to NF-κB mask its NLS so that IκB–NF-κB complexes resides in the cytoplasm. The extracellular presence of a remarkable variety of external stimuli including certain bacterial and viral products, several cytokines (Section 34-4B), phorbol esters (*p.* 1290), and oxidative or physical stress, results in IκBs being rapidly phosphorylated and proteolytically degraded in the proteosome (Section 30-6B and Section 30-6a of this Supplement). The newly liberated NF-κB is thereupon translocated to the nucleus, where it mediates transcriptional initiation by binding to 10-bp DNA sequences of the sort GGGRNNYYCC (where N is any nucleotide, R is a purine, and Y is a pyrimidine). Such NF-κB binding sites have been identified as important regulatory motifs in numerous genes such as those encoding certain cytokines and cytokine receptors, Class I MHC proteins (Section 34-2E), and certain viral proteins including those of HIV. Additional specificity may be achieved through the synergistic interaction of NF-κB with other DNA-bound transcription factors such as Sp1 and Jun/AP-1 (Section 34-4B).

In an alternative mode of NF-κB activation, p50 is synthesized as the N-terminal domain of **p105,** a protein whose C-terminal domain is an IκB. The IκB domain of p105 prevents both the nuclear localization and the DNA binding of p105 as well as other RHR-containing proteins. The above external stimuli also accelerate the proteolytic processing of p105 to yield free NF-κB and the IκB-containing C-terminal domain of p105, which as above, is phosphorylated and proteolytically degraded.

Two X-ray structures of NF-κB p50 in complex with its target DNA were simultaneously reported: that of mouse p50 (residues 39-364) with the 11-nt DNA TGGGAATTCCC (which forms a 10-bp duplex with single T overhangs at each end), the sequence of the symmetrized consensus κB site; and that of human p50 (residues 2-366) with the 19-nt DNA AGAT<u>GGGGAATCCCC</u>TAGA [which forms a central 11-bp duplex (underlined) that has a central A·A mismatch], the sequence of an MHC Class I enhancer site to which p50 preferentially binds. The RHRs in the two proteins (mouse residues 40-363 and human residues 43-366) have nearly identical sequences.

The structures of both complexes bear a remarkable resemblance to a butterfly in which the body is formed by the duplex DNA and each outspread wing is formed by a subunit of the p50 dimer. The mouse and human proteins have closely similar structures consisting of two domains, with the C-terminal domains forming the dimerization interface and both domains interacting with the DNA. Both these N- and C-terminal domains have immunoglobulin-like folds (Fig. 34-34) and interact with the DNA exclusively through 10 loops (5 from each subunit) that connect β strands in a manner reminiscent of the way that p53, which has a core structure similar to that of NF-κB's N-terminal domain, interacts with its target DNA (Fig. 33-87). The resulting DNA-binding surface is much more extensive than those of other transcription factors, which accounts for the unusually high affinity of NF-κB for its target DNA. This also explains the inability of deletion mutagenesis to localize p50's DNA-binding region, since changes anywhere in its structure are likely to affect the disposition of its DNA-binding loops. The B-DNAs in both complexes have quite deep major grooves and rather narrow minor grooves around their central bp's.

Mutational changes in the Dorsal protein at positions corresponding to residues Thr 256 and Ala 257 of mouse NF-κB disrupt its interactions with the dorsal-specific IκB named **Cactus**. These residues are located above p50's dimer interface facing away from the DNA-binding surface. Mutational studies have also demonstrated that at least some IκB proteins bind to the C-terminal 5 residues forming the NLS, which in both the mouse and human proteins are part of a longer disordered C-terminal segment. In fact, the last ordered residue in mouse p50, Glu 350, is close to residues 256 and 257.

Mouse p50 is bound to a 10-bp DNA segment, whereas human p50 is bound to an 11-bp DNA segment. Thus, the major difference between the two complexes is in the half-site spacing of the two DNAs. Comparison of the two structures indicates that this difference is accommodated in part by conformational changes at the polypeptide segment linking the N- and C-terminal domains of p50 (residues 241–247 in mouse and 244-250 in human) that close the "jaws" of the two N-terminal domains of the mouse p50 dimer more tightly about the DNA than those of human p50. The remaining significant differences between the two structures is taken up by the DNA, which in the mouse complex is essentially straight but in the human complex is bent by ~15° across the complex's 2-fold axis towards the dimerized C-terminal domains. Through these conformational shifts, p50 maintains many of the same contacts with the DNA in both complexes.

Chytil, M. and Verdine, G.L., The rel family of eukaryotic transcription factors, *Curr. Opin. Struct. Biol.* **6,** 91–100 (1996).

Ghosh, G., Van Duyne, G., Ghosh, S., and Sigler, P.B., Structure of NF-κB p50 homodimer bound to a κB site, *Nature* **373,** 303–310 (1995).

Kuriyan, J. and Thanos, D., Structure of the NF-κB transcription factor: a holistic interaction with DNA, *Structure* **3,** 155–141 (1995).

Müller, C.W., Rey, F.A., Sodeoka, M., Verdine, G.L., and Harrison, S.C., Structure of the NF-κB p50 homodimer bound to DNA, *Nature* **373,** 317–310 (1995).

Müller, C.W., Rey, F.A., and Harrison, S.C., Comparison of the two different DNA-binding modes of the NF-κB p50 homodimer, *Nature Struct. Biol.* **3,** 224–227 (1996).

(c) Additional References

Beelman, C.A. and Parker, R., Degradation of mRNA in eukaryotes, *Cell* **81,** 179–183 (1995).

Berg, J.M. and Shi, Y., The galvanization of biology: a growing appreciation for the roles of zinc, *Science* **271,** 1081–1085 (1996).

Berget, S.M., Exon recognition in vertebrate splicing, *J. Biol. Chem.* **270,** 2411–2414 (1995).

Burley, S.K., The TATA box binding protein, *Curr. Opin. Struct. Biol.* **6,** 69–75 (1996).

Chen, C.-Y.A. and Shyu, A.-B., AU-rich elements: characterization and importance in mRNA degradation, *Trends Biochem. Sci.* **20,** 465–470 (1995).

Davis, L.I. The nuclear pore complex, *Annu. Rev. Biochem.* **64,** 865–896 (1995).

Loo, S. and Rine, J., Silencing and heritable domains of gene expression, *Annu. Rev. Cell Biol.* **11,** 519–548 (1995).

Glover, J.N.M. and Harrison, S.C., Crystal structure of the heterodimeric bZIP transcription factor c-Fos–c-Jun bound to DNA, *Nature* **373,** 257–261 (1995).

Goodrich, J.A., Cutler, G., and Tjian, R., Contacts in context: promoter specificity and macromolecular interactions in transcription, *Cell* **84,** 825–830 (1996).

Guarente, L., Transcriptional coactivators in yeast and beyond, *Trends Biochem. Sci.* **20,** 517–521 (1995).

Izaurralde, E. and Mattaj, I.W., RNA export, *Cell* **81,** 153–159 (1995).

Keller, W., König, P., and Richmond, T., Crystal structure of a bZIP/DNA complex at 2.2 Å: Determinants of DNA specific recognition, *J. Mol. Biol.* **254,** 657–667 (1995).

Kodadek, T., From carpet bombing to cruise missiles: the 'second-order' mechanisms used by transcription factors to ensure specific DNA binding *in vivo, Chemistry & Biology* **2,** 267–279 (1995).

Koleske, A.J. and Young, R.A., The RNA polymerase II holoenzyme and its implications for gene regulation, *Trends Biochem. Sci.* **20,** 113–116 (1995).

Love, J.J., Li. X., Case, D.A., Giese, K., Grosschedl, R., and Wright, P.E., Structural basis for DNA bending by the architectural transcription factor LEF-1, *Nature* **376,** 791–795 (1995).

McCarthy, J.E.G. and Kollmus, H., Cytoplasmic mRNA–protein interactions in eukaryotic gene expression, *Trends Biochem. Sci.* **20,** 191–198 (1995).

Ogata, K., Morikawa, S., Nakamura, H., Sekikawa, A., Inoue, T., Kanai, H., Sarai, A., Ishii, S., and Nishimura, Y., Solution structure of a specific DNA complex of the Myb DNA-binding domain with cooperative recognition helices, *Cell* **79,** 639–648 (1994).

Orkin, S.H., Transcription factors and hematopoietic development, *J. Biol. Chem.* **270,** 4955–4958 (1995).

Penny, G.D., Kay, G.F., Sheardown, S.A, Rastan, S., and Brockdorff, N., Requirement for *Xist* in X chromosome inactivation, *Nature* **379,** 131–137 (1996).

Sassone-Corsi, P., Transcription factors responsive to cAMP, *Annu. Rev. Cell Biol.* **11,** 355–377 (1995).

Shaw, P.J. and Jordan, E.G., The nucleolus, *Annu. Rev. Cell Biol.* **11,** 93–121 (1995).

St Johnston, D., The intracellular localization of messenger RNAs, *Cell* **81,** 161–170 (1995).

Studitsky, V.M., Clark, D.J. and Felsenfeld, G., Overcoming a nucleosomal barrier to transcription, *Cell* **83,** 19–27 (1995).

Stuhl, K., Chromatin structure and RNA polymerase II connection: Implications for transcription, *Cell* **84,** 179–182 (1996).

Towle, H.C., Metabolic regulation of gene transcription in mammals, *J. Biol. Chem.* **270,** 23235–23238 (1995).

Werner, M.H., Clore, G.M., Fisher, C.L., Fisher, R.J., Trinh, L., Shiloach, J., and Gronenborn, A.M., The solution structure of the human ETS1–DNA complex reveals a novel mode of binding and true side chain intercalation, *Cell* **83,** 761–771 (1995).

Wu, C., Heat shock transcription factors, *Annu. Rev. Cell Biol.* **11,** 441–469 (1995).

Zawel, L., and Reinberg, D., Common themes in assembly and function of eukaryotic transcription complexes, *Annu. Rev. Biochem.* **64,** 533–561 (1995).

4. CELL DIFFERENTIATION

(a) Expression of the *Drosophila eyeless* Gene Induces the Ectopic Formation of Eyes

Mutations in the *Drosophila eyeless* (*ey*) gene, first described in 1915, result in flies whose compound eyes are reduced in size or completely absent. The *ey* gene encodes a transcription factor that contains both a homeodomain and a so-called **paired** domain. The expression of *ey* is first detected in the embryonic nervous system and later in the embryonic primordia of the eye. In subsequent larval stages, it is expressed in the developing eye imaginal discs. Mutant forms of four other *Drosophila* genes that have similar phenotypes do not affect the expression of the *ey* gene, which indicates that *ey* acts before these other genes. These observations led to the suggestion that the *ey* gene is the master control gene for eye development.

Genetic engineering studies have confirmed this hypothesis. Through the targeted expression of *ey* cDNA in various imaginal disc primordia of *Drosophila,* ectopic (inappropriately positioned) compound eyes were induced to form on the wings, legs, and antennae of various flies. Moreover, in many cases, these eyes appeared morphologically normal in that they consisted of fully differentiated ommatidia (the simple eye elements that form a compound eye) with a complete set of photoreceptor cells that appear to be electrically active when illuminated (although it is unknown if the flies could see with these ectopic eyes, that is, whether the eyes made appropriate neural connections to the brain).

The mouse *Small eye* (*Sey* or *Pax*-6) gene and the human *Aniridia* gene are closely similar in sequence to the *Drosophila ey* gene and are similarly expressed during morphogenesis. Mice with mutations in one of their two *Sey* genes have underdeveloped eyes, whereas those with mutations in both *Sey* genes are eyeless. Similarly, humans that are heterozygotes for a defective *Aniridia* gene have defects in their iris, lens, cornea, and retina. Evidently, the *ey, Sey,* and *Aniridia* genes all function as master control genes for eye formation in their respective organisms, a surprising result considering the enormous morphological differences between insect and mammalian eyes. Thus, despite the 500 million years since the divergence of insects and mammals, their developmental control mechanisms appear to be closely related.

Halder, G., Callaerts, P., and Gehring, W.J., Induction of ectopic eyes by targeted expression of the *eyeless* gene in *Drosophila, Science* **267,** 1788–1792 (1995).

(b) Bicoid Protein Binds to and Translationally Represses the Translation of *caudal* mRNA

The mRNA encoding the *Drosophila* Bicoid protein, a homeodomain-containing product of the maternal-effect gene *bicoid* (*bcd*), is laid down during oogenesis at the egg's anterior pole. After fertilization, *bcd* mRNA is translated giving rise to Bicoid protein in an anterior-to-posterior gradient that, in turn, activates several segmentation genes in a concentration-dependent manner (Fig. 33-70*a,b*). Shortly after the Bicoid protein gradient forms, an opposing posterior-to-anterior gradient of the protein **Caudal** (Latin: *cauda*, tail), which also contains a homeodomain, forms under the control of Bicoid protein. The observation that the anterior expression of Caudal protein causes the deletion of head and thoracic segments indicates that the Bicoid-mediated regulation of Caudal is necessary for normal embryonic patterning. How does Bicoid protein control the opposite gradient of Caudal protein?

Maternally supplied *caudal* (*cad*) mRNA is evenly distributed in the early embryo. Thus, the observation that mutant embryos lacking Bicoid activity fail to form a Caudal gradient suggests that Bicoid protein acts to repress the translation of *cad* mRNA. Experiments in two laboratories have demonstrated that the homeodomain-containing segment of Bicoid protein binds to a 342-nt segment of the 940-nt 3' untranslated region (**UTR**) of *cad* mRNA. The use of reporter genes has shown that the interaction between Bicoid protein and this Bicoid-binding region of *cad* mRNA indeed represses translational initiation at the 5' end of the mRNA. Thus Bicoid protein acts both as a transcriptional activator that binds to the upstream segments of a variety of DNAs and as a translational repressor that binds to the *cad* mRNA's 3' UTR.

Dubnau, J. and Struhl, G., RNA recognition and translational regulation by a homeodomain protein, *Nature* **379,** 694–699 (1996).

Rivera-Pomar, R., Niessing, D., Schmidt-Ott, U., Gehring, W.J., and Jäckle, H., RNA binding and translational suppression by bicoid, *Nature* **379,** 746–749 (1996).

(c) Structures of Cyclin, Cks Proteins, and Their Complexes with CDK2

Progression through the cell cycle is mediated by the activation of the appropriate members of the cyclin-dependent protein kinase (Cdk or **CDK**) family of Ser/Thr protein kinases. This occurs, in part, through their binding of specific cyclins, proteins that were so-named because of their accumulation and abrupt ubiquitin-mediated proteolytic destruction in particular portions of the cell cycle (Fig. 33-78). The cyclins form a diverse family of proteins that share moderate homology (30-50% similarity) over a ~100-residue segment, the so-called **cyclin box;** eight types of cyclins (A-H) have been characterized in higher eukaryotes. CDKs, in contrast, are highly homologous (40-75% similarity) and contain a conserved core of ~300 residues. In fact, comparisons of the X-ray structures of several eukaryotic protein kinases (Section 17-3d of this Supplement), including human CDK2 (Fig. 33-79) and mouse cAMP-dependent protein kinase (cAPK; Fig. 17-14), reveal that all of them have a similar tertiary structure, although they also have significant differences.

The presence of **cyclin A** is required for progression through S phase of the cell cycle (Fig. 33-62) and for entry into mitosis. The X-ray structure of the C-terminal portion of bovine cyclin A (residues 170- 432) reveals that this protein fragment folds into a compact globule that contains 12 α helices (and no β sheets), of which the cyclin box (residues 208-303) forms helices 2 to 5. Unexpectedly, that segment of the structure encompassing helices 6 to 11 (residues 309-399) forms a closely similar structure, despite the only 12% identical sequences of these two segments.

The X-ray structure of human CDK2 in complex with ATP and the corresponding fragment of human cyclin A (residues 173-432) indicates that cyclin A binds to one side of CDK2's catalytic cleft (its right side in Fig. 33-79), where it interacts with both lobes of the CDK2 to form an extensive and continuous protein–protein interface. Cyclin A does not undergo significant conformational changes on binding CDK2. In contrast, the binding of cyclin A causes CDK2 to undergo significant conformational shifts in the region around its catalytic cleft. In particular, the α1 helix of CDK2 (see Fig. 33-79), which contains the PSTAIRE sequence motif (using the one-letter amino acid code; Table 4-1) characteristic of the CDK family, rotates about its axis and moves several Å into the catalytic cleft relative to its position in free CDK2, where it contacts the cyclin box segment of cyclin A. This movement brings Glu 51 from its solvent-exposed position outside the catalytic cleft of free CDK2 to a position inside the catalytic cleft where it forms a salt bridge with Lys 33, which in free CDK2 instead forms a salt bridge to Asp 145. These three side chains (Lys 33, Glu 51, and Asp 145), which are conserved in all eukaryotic protein kinases, participate in ATP phosphate coordination and Mg^{2+} ion coordination. Their conformational reorientation on cyclin A binding apparently places them in a catalytically active arrangement.

The binding of cyclin A also induces CDK2's T-loop (residues 152–170; L12 in Fig. 33-79), which blocks the entrance to the catalytic cleft in free CDK2, to undergo extensive conformational changes. This results in positional shifts of up to 21 Å, such that the T-loop, which now also contacts the cyclin box, adopts a backbone conformation that closely resembles that of the analogous region of cAPK (which is thought to have a catalytically active conformation). These movements greatly increase the access of a protein substrate to the ATP bound in the catalytic cleft. The movement of the T-loop also better exposes its component Thr 160, which must be phosphorylated by Cdk-activating kinase for full CDK2 activity.

In addition to both positive and negative control by phosphorylation and by the binding of cyclins, CDKs bind a class of regulatory proteins represented by **Cks1** (for *Cdc28 kinase subunit*) in *Saccharomyces cerevisiae* and by **p13^{suc1}** (for *supressor of cdc2 temperature-sensitive mutations*) in *Schizosaccharomyces pombe* (fission yeast). The observation that two human homologs of Cks1, **CksHs1** and **CksHs2,** can be functionally substituted for Cks1 in *S. cerevisiae* indicates that these proteins are both structurally and functionally conserved. Although genetic analysis indicates that yeast Cks proteins have an essential role in cell cycle progression, they are not substrates of any kinase nor do they possess catalytic activity. Indeed, their precise biological function is unclear.

The X-ray structures of CksHs1, CksHs2, and two crystal forms of p13^{suc1} have been determined. Each of these ~90-residue proteins have quite similar structures that consist mainly of a 4-stranded antiparallel β sheet flanked by two short helices. CksHs1 and p13^{suc1} in one of its structures crystallize as monomers. However, in its other structure, p13^{suc1} crystallizes as a dimer, whereas CksHs2 crystallizes as a hexamer that may be considered to be a trimer of dimers. In each of these dimers, the C-terminal β strand is exchanged between the two subunits, although, as the comparison of the two p13^{suc1} structures reveals, this does not greatly perturb the conformation of these subunits.

The X-ray structure of CksHs1 in complex with CDK2 reveals that a single subunit of CksHs1 binds, via all four of its β strands, to the C-terminal domain of CDK2 (bottom of Fig. 33-79) so as to engage its L14 loop and α5 helix. The binding of CksHs1 does not appear to affect the structure of CDK2, indicating that Cks proteins are unlikely to regulate CDK proteins by inducing a conformational change. Moreover, the 26-Å distance between CksHs1 and CDK2 Tyr 15 does not support the previously made hypothesis that the formation of the CksHs1 complex interferes with the dephosphorylation of CDK2 Tyr 15 and hence the activation of CDK2. Superposition of the β-interchanged p34^{suc1} or CksHs2 dimers on CksHs1 bound to CDK2 reveals several steric clashes between the CDK2 N-terminal lobe and the adjacent subunit of the Cks dimer, which suggests that the β-exchanged Cks dimer cannot bind to CDK2. Thus the dimerization of Cks proteins under appropriate conditions associated with cell cycle transitions may provide a mechanism for limiting the binding of Cks proteins to CDKs. Comparison of the CDK2–CksHs1 and CDK2–cyclin A complexes indicates that CksHs1 and cyclin A can simultaneously bind to CDK2 as has been shown to occur in solution. The sequence-conserved pocket on the surface of CksHs1 to which the phosphate analog vandadate is bound in the structure of CksHs1 alone, is exposed on the same side of the CDK2–CksHs1 complex as is the CDK2 catalytic site. CksHs1 thereby extends the recognition surface flanking the CDK2 catalytic site, which suggests that Cks binding targets the CDK to substrates or to other phosphorylated proteins.

Arvai, A.S., Bourne, Y., Hickey, M.J., and Tainer, J.A., Crystal structure of the human cell cycle protein CksHs1: Single domain fold with similarity to kinase N-lobe domain, *J Mol. Biol.* **249,** 835–842 (1995).

Bourne, Y., Watson, M.H., Hickey, M.J., Holmes, W., Rocque, W., Reed, S.I., and Tainer, J.A., Crystal structure and mutational analysis of the human CDK2 kinase complex with cell cycle–regulatory protein CksHs1, *Cell* **84**, 863–874 (1996).

Brown, N.R., Noble, M.E.M., Endicott, J.A., Garman, E.F., Wakatsuki, S., Mitchell, E., Rasmussen, B., Hunt, T., and Johnson, L.N., The crystal structure of cyclin A, *Structure* **3,** 1235–1247 (1995).

Endicott, J.A., Noble, M.E., Garman, E.F., Brown, N., Rasmussen, B., Nurse, P., and Johnson, L.N., The crystal structure of p13^{suc1}, a p34^{cdc2}-interacting cell cycle control protein, *EMBO J.* **14,** 1004–10014 (1995).

Jeffrey, P.D., Russo, A.A., Polyak, K., Gibbs, E., Hurwitz, J., Massagué, J., and Pavletich, N.C., Mechanism of CDK activation revealed by the structure of a cyclinA–CDK2 complex, *Nature* **376,** 313–320 (1995).

Khazanovich, N., Bateman, K.S., Chernaia, M., Michalak, M., and James, M.N.G., Crystal structure of the yeast cell-cycle control protein p13^{suc1} in a strand-exchanged dimer, *Structure* **4,** 299–309 (1996).

MacLachlan, T.K., Sang, N., and Giordano, A., Cyclins, cyclin-dependent kinases and Cdk inhibitors: implications in cell cycle control and cancer, *Crit. Rev. Euk. Gen Exp.* **5,** 127–156 (1995).

Morgan, D.O., Principles of CDK regulation, *Nature* **374,** 131–134 (1995).

Parge, H.E., Arvai, A.S., Murtari, D.J., Reed, S.I. and Tainer. J.A., Human CksHs2 atomic structure: a role for its hexameric assembly in cell cycle control, *Science* **262,** 387–395 (1993).

Radzio-Andzelm, E., Lew, J., and Taylor, S., Bound to activate: conformational consequences of cyclin binding to CDK2, *Structure* **3,** 1135–1141 (1995).

(d) Additional References

Averof, M. and Akam, M., Hox genes an the diversification of insect and crustacean body plans, *Nature* **376**, 420–423 (1995).

Campisi, J., Replicative senescence: An old lives tale, *Cell* **84,** 497–500 (1996).

Carroll, S.B., Homeotic genes and the evolution of arthropods and chordates, *Nature* **376,** 479–485 (1995).

Chen, P.-L. Riley, D.J., and Lee, W.-H., The retinoblastoma protein as a fundamental mediator of growth and differentiation signals, *Crit. Rev. Euk. Gene Exp.* **5,** 79–95 (1995).

Chernova, O.B., Chernov, M.V., Agarwal, M.L., Taylor, W.R., and Stark, G.R., The role of p53 in regulating genomic stability when DNA and RNA synthesis are inhibited, *Trends Biochem. Sci.* **20,** 431–434 (1995).

Curtis, D., Lehmann, R., and Zamore, P.D., Translational regulation in development, *Cell* **81,** 171–178 (1995).

Enoch, T. and Norbury, C., Cellular responses to DNA damage: cell-cycle checkpoints, apoptosis and the roles of p53 and ATM, *Trends Biochem. Sci.* **20,** 426–430 (1995).

Hartwell, L.H. and Kastan, M.B., Cell cycle control and cancer, *Science* **266,** 1821–1828 (1994).

Jeffrey, P.D., Gorina, S., and Pavletich, N., Crystal structure of the tetramerization domain of the p53 tumor suppressor at 1.7 Angstroms, *Science* **267,** 1498–1502 (1995); *and* Clore, M.G., Ernst, J., Clubb, R.,Omichinski, J.G., Kennedy, W.M.P., Sakaguchi, K., Appella, E., and Gronenborn, A.M., Refined solution structure of the oligomerization domain of the tumour suppressor p53, *Nature Struct. Biol.* **2,** 321–332 (1995).

Kamb, A., Cell-cycle regulators and cancer, *Trends Genet.* **11,** 136–140 (1995).

Li, T., Stark, M.R., Johnson, A.D., and Wolberger, C., Crystal structure of the MATa1/Matα2 homeodomain heterodimer bound to DNA, *Science* **270,** 262–269 (1995); *and* Jin, Y., Mead, J., Li, T., Wolberger, C., and Vershon, A.K., Altered DNA recognition and bending by insertions in the α2 tail of the yeast a1/α2 homeodomain heterodimer, *Science* **270,** 290–293 (1995).

Murray, A., Cyclin ubiquitination: the destructive end of mitosis, *Cell* **81,**149–152 (1995).

Weinberg, R.A., The retinoblastoma protein and cell cycle control, *Cell* **81,** 323–330 (1995).

Wolberger, C., Homeodomain interactions, *Curr. Opin. Struct. Biol.* **6,** 62–68 (1996).

Xu, W., Rould, M.A., Jun, S., Desplan, C., and Pabo, C.O., Crystal structure of a Paired domain–DNA complex at 2.5 Å reveals structural basis for Pax developmental mutations, *Cell* **80,** 639–650 (1995).

Yu, C.-E., et al, Positional cloning of Werner's syndrome gene, *Science* **272,** 258–262 (1996). [The gene for a premature aging syndrome encodes a 1432-residue protein similar to known DNA helicases.]

Chapter 34

MOLECULAR PHYSIOLOGY

1. BLOOD CLOTTING

(a) Clotting Factors

Banner, D.W., D'Arcy, A., Chène, C., Winkler, F.K., Guha, A., Konigsberg, W.H., Nemerson, Y., and Kirchhofer, D., The crystal structure of the complex of blood coagulation factor VIIa with tissue coagulation factor, *Nature* **380,** 41–46 (1996). [The complex whose formation initiates blood coagulation.]

Dowd, P., Hershline, R., Ham, S.W., Naganathan, S., Vitamin K and energy transduction: a base strength amplification mechanism, *Science* **269,** 1684–1691 (1995).

Gibbs, C.S., et al, Conversion of thrombin into an anticoagulant by protein engineering, *Nature* **378,** 413–416 (1995).

Rao, Z., Handford, P., Mayhew, M., Knott, V., Brownlee, G.G., and Stuart, D., The structure of Ca^{2+}-binding epidermal growth factor-like domain: its role in protein–protein interactions, *Cell* **82,** 131–141 (1995). [The structure of an EGF-like domain from Factor IX.]

Stubbs, M.T. and Bode, W., The clot thickens: clues provided by the thrombin structure, *Trends Biochem. Sci.* **20,** 23–28 (1995); *and* Coagulation factors and their inhibitors, *Curr. Opin. Struct. Biol.* **4,** 823–832 (1994). [Reviews.]

(b) Control of Clotting

Carrell, R.W., Stein, P.E., Fermi, G., and Wardell, M.R., Biological implications of a 3 Å structure of dimeric antithrombin, *Structure* **2,** 257–270 (1995).

(c) Clot Lysis

Spraggon, G., Phillips, C., Nowak, U.K., Ponting, C.P., Saunders, D., Dobson, C.M., Stuart, D.I., and E.Y. Jones, The crystal structure of the catalytic domain of human urokinase-type plasminogen activator, *Structure* **3,** 681–691 (1995). [The first known structure of a fibrinolytic enzyme.]

2. IMMUNITY

(a) Antibody Structures

Barré, S., Greenberg, R.S., Flajnik, M.F., and Chothia, C., Structural conservation of hypervariable regions in immunoglobulin evolution, *Nature Struct. Biol.* **1,** 915–920 (1994).

Burmeister, W.P., Gastinel, L.N., Simister, N.E., Blum, M.L., and Bjorkamna, P.J., Crystal structure at 2.2 Å resolution of MHC-related neonatal Fc receptor, *Nature* **372,** 336–343 (1994); *and* Burmeister, W.P., Huber, A.H., and Bjorkman, P.J., Crystal structure of the complex of rat neonatal Fc receptor with Fc, *Nature* **372,** 379–383 (1994). [Neonatal Fc receptor transports maternal IgG from milk to the blood stream of the newborn.]

Davies, D.R. and Cohen, G.H., Interactions of protein antigens with antibodies, *Proc. Natl. Acad. Sci.* **93,** 7–12 (1996).

Stanfield, R.L. and Wilson, I.A., Protein–peptide interactions, *Curr. Opin. Struct. Biol.* **5,** 103–113 (1995); *and* Wilson, I.A. and Stanfield, R.L., Anitbody–antigen interactions: new structures and new conformational changes, *Curr. Opin. Struct. Biol.* **4,** 847–867 (1994).

(b) Catalytic Antibodies – Abzymes

Hilvert, D., Catalytic antibodies, *Curr. Opin. Struct. Biol.* **4,** 612–617 (1994).

Jacobsen, J.R. and Schultz, P.G., The scope of antibody catalysis, *Curr. Opin. Struct. Biol.* **5,** 818–824 (1995).

Schultz, P.G. and Lerner, R.A., From molecular diversity to catalysis: lessons from the immune system, *Science* **269,** 1835–1842 (1995).

(c) Generation of Antibody Diversity

Hengtschläger, M., Maizels, N., and Leung, H., Targeting and regulation of immunoglobulin gene somatic hypermutation and isotype switch recombination, *Prog. Nucl. Res. Mol. Biol.* **50,** 67–99 (1995).

Jackson, S.P. and Jeggo, P.A., DNA double-strand break repair and V(D)J recombination: involvement of DNA-PK, *Trends Biochem. Sci.* **20,** 412–415 (1995). [Discusses the action of a DNA-dependent protein kinase (DNA-PK).]

van Gent, D.C., Mizuuchi, K., and Gellert, M., Similarities between initiation of V(D)J recombination and retroviral integration, *Science* **271,** 1592–1594 (1996); McBane, J.F., van Gent, D.C., Ramsden, D.A., Romeo, C., Cuomo, C.A., Gellert, M., and Oettinger, M.A., Cleavage at a V(D)J recombination signal requires only RAG1 and RAG2 proteins and occurs in two steps, *Cell* **83,** 387–395 (1995); *and* van Gent, D.C., McBlane, J.F., Ramsden, D.A., Sadofsky, M.J., Hesse, J.E., and Gellert, M., Initiation of V(D)J recombination in a cell-free system, *Cell* **81,** 925–934 (1995).

(d) *T* Cell Receptors

Bentley, G.A., Boulot, G., Karjalainen, K., and Mariuzza, R.A., Crystal structure of the β chain of a T cell antigen receptor, *Science* **267,** 1984–1987 (1995).

Bluestone, J.A., Khattri, R., Sciammas, R., and Sperling, A.I., TCRγδ cells: A specialized T-cell subset in the immune system, *Annu. Rev. Cell Dev. Biol.* **11,** 307–3353 (1995).

Fields, B.A., Ober, B., Malchiodi, E.L., Lebedeva, M.I., Braden, B.C., Ysern, X., Kim, J.-K., Shao, X., Ward, E.S., and Mariuzza, R.A., Crystal structure of the V_α domain of a T cell antigen receptor, *Science* **270,** 1821–1824 (1995).

Hatada, M.H., et al., Molecular basis for interaction of the protein tyrosine kinase ZAP-70 with T-cell receptor, *Nature* **377,** 32–38 (1995). [Shows the interaction of the tandem SH2 domains of human ZAP-70 in complex with a peptide derived from the ζ subunit of the *T* cell receptor to yield an interface that forms a phosphotyrosine binding site.]

(e) The Major Histocompatibility Complex

Bjorkman, P.J. and Burmeister, W.P., Structures of two classes of MHC molecules elucidated: crucial differences and similarities, *Curr. Opin. Struct. Biol.* **4,** 852 (1994).

Ghosh, P., Amaya, M., Mellins, E., and Wiley, D.C., The structure of an intermediate in class II MHC maturation: CLIP bound to HLA-DR3, *Nature* **378,** 457–462 (1995).

Heemels, M.T. and Ploegh, H., Generation, translocation, and presentation of MHC Class I-restricted peptide, *Annu. Rev. Biochem.* **64,** 463–491 (1995).

Madden, D.R., The three-dimensional structure of peptide-MHC complexes, *Annu. Rev. Immunol.* **13,** 587–622 (1995).

Wolf, P.R. and Ploegh, H.L., How MHC ClassII molecules acquire peptide cargo, *Annu. Rev. Cell Dev. Biol.* **11,** 267–306 (1995).

(f) The Complement System

Dodds, A.W., Ren, X.-D., Willis, A.C., and Law, S.K.A., The reaction mechanism of the internal thioester in the human complement component C4, *Nature* **379,** 177–179 (1996).

3. MOTILITY: MUSCLES, CILIA, AND FLAGELLA

(a) Structure of Muscle

Campbell, K.P., Three muscular dystrophies: Loss of cytoskeletal–extracellular matrix linkage, *Cell* **80,** 675–679 (1995). [A review.]

Fisher, A.J., Smith, C.A., Thoden, J.B., Smith, R., Sutoh, K., Holden, H.M., and Rayment, I., X-ray structure of the myosin motor domain of *Dictostelium discoideum* complexed with MgADP·BeF$_x$ and MgADP·AlF$_4^-$; *and* Smith, C.A. and Rayment, I., X-ray structure of the magnesium(II)–pyrophosphate complex of the truncated head of *Dictostelium discoideum* myosin at 2.7 Å resolution, *Biochemistry* **34,** 8960–8972 *and* 8973–8981 (1995).

Improta, S., Politou, A.S., and Pastore, A., Immunoglobulin-like modules from titan I-band: extensible components of muscle elasticity, *Structure* **4,** 323–337 (1996). [The NMR structure of an Ig module from the I-band.]

Labeit, S. and Kolmerer, B., Titans: giant proteins in charge of muscle ultrastructure and elasticity, *Science* **270,** 293–296 (1995). [The DNA sequence of human cardiac titan and its implied amino acid sequence.]

Lorenz, M., Poole, K.J.V., Popp, D., Rosenbaum, G., and Holmes, K.C., An atomic model of unregulated thin filament obtained by X-ray fiber diffraction on oriented actin-tropomyosin gels,

J. Mol. Biol. **246,** 108–119 (1995).

Milligan, R.A., Protein–protein interactions in the rigor actomyosin complex, *Proc. Natl. Acad. Sci.* **93,** 21–26 (1996).

Mooseker, M.S. and Cheney, R.E., Unconventional myosins, *Annu. Rev. Cell Dev. Biol.* **11,** 633–675 (1995).

Rupple, K.M., Lorenz, M., and Spudich, J.A., Myosin structure/function: a combined mutagenesis-crystallography approach, *Curr. Opin. Struct. Biol.* **5,** 181–186 (1995).

Schafer, D.A. and Cooper, J.A., Control of actin assembly at filament ends, *Annu. Rev. Cell Dev. Biol.* **11,** 497–518 (1995).

(b) Mechanism and Control of Muscle Contraction

Egelman, E.H., and Orlova, A., New insights into actin filament dynamics, *Curr. Opin. Struct. Biol.* **5,** 172–180 (1995).

Gagné, S.M., Tsuda, S., Li, M.X., Smillie, L.B., and Sykes, B.D., Structures of the troponin C regulatory domains in the apo and calcium-saturated states, *Nature Struct. Biol.* **2,** 784–789 (1995).

Houdussse, A. and Cohen, C., Structure of the regulatory domain of scallop myosin at 2 Å resolution: implications for regulation, *Structure* **4,** 21–32 (1996).

Spudich, J.A., How molecular motors work, *Nature* **372,** 515–518 (1994).

Whittaker, M., Wilson-Kubalek, E.M., Smith, J.E., Faust, L., Milligan, R.A., and Sweeney, H.L., A 35-Å movement of smooth muscle myosin on ADP release; *and* Jontes, J.D., Wilson-Kubalek, E.M., and Milligan, R.A., A 32° tail swing in brush border myosin I on ADP release, *Nature* **378,** 748–751 *and* 751–753 (1995).

(c) Tubulin and Kinesin

Hirose, K., Lockhart, A., Cross, R.A., and Amos, L.A., Nucleotide-dependent angular change in kinesin motor domain bound to tubulin, *Nature* **376,** 277–279 (1995).

Hoenger, A., Sablin, E.P., Vale, R.D., Fletterick, R.J., and Milligan, R.A., Three-dimensional structure of a tubulin–motor protein complex, *Nature* **376,** 271–274 (1995).

Kull, F.J., Sablin, E.P., Lau, R., Fletterick, R.J., and Vale, R.D., Crystal structure of the kinesin motor domain reveals a structural similarity to myosin; *and* Sablin, E.P., Kull, F.J., Cooke, R., Vale, R.D., and Fletterick, R.J., Crystal structure of the motor domain of the kinesin-related motor ncd, *Nature* **380,** 550–554 *and* 555–559 (1996).

Nogales, E., Wolf, S.G., Khan, I.A., Ludueña, R.F., and Downing, K.H., Structure of tubulin at

6.5 Å and location of the taxol-binding site, *Nature* **375,** 424–427 (1995).

Zheng, Y., Wong, M.L., Alberts, B., and Mitchison, T., Nucleation of microtubule assembly by a γ-tubulin-containing ring complex; *and* Moritz, M., Braunfeld, M.B., Sedat, J.W., Alberts, B., and Agard, D.A., Microtubule nucleation by γ-tubulin-containing rings in the centrosome, *Nature* **378,** 578–583 *and* 638–640 (1995).

(d) Bacterial Flagella

DeRosier, D.J., Spinning tails, *Curr. Opin. Struct. Biol.* **5,** 187–193 (1995). [Discusses the structure of the bacterial flagellar rotary motor.]

Mimori, Y., Yamashita, I., Murata, K., Fujiyoshi, Y., Yonekura, K., Toyoshima, C., and Namba, K., The structure of the R-type straight flagellar filament of *Salmonella* at 9 Å resolution by electron cryomicroscopy, *J. Mol. Biol.* **249,** 69–87 (1995).

Morgan, D.G., Owen, C., Melanson, L.A., and DeRosier, D.J., Structure of bacterial flagellar filaments at 11 Å resolution: packing of α-helices, *J. Mol. Biol.* **249,** 88–110 (1995).

Shapiro, L., The bacterial flagellum: from genetic network to complex architecture, *Cell* **80,** 525–527 (1995).

4. BIOCHEMICAL COMMUNICATIONS: HORMONES AND NEUROTRANSMISSION

(a) Signal Transduction – Miscellaneous

Cell **80,** 179–278 (1995). [A series of authoritative reviews on signal transduction.]

Hilgenfeld, R., Regulatory GTPases, *Curr. Opin. Struct. Biol.* **5,** 810–817 (1995).

Schwartz, M.A., Schaller, M.D., and Ginsberg, M.H., Integrins: emerging paradigms of signal transduction, *Annu. Rev. Cell Dev. Biol.* **11,** 549–599 (1995).

(b) Growth Factors and Their Receptors

Heaney, M.L. and Golde, D.W., Soluble cytokine receptors, *Blood* **87,** 847–857 (1996).

Somers, W., Ultsch, M., De Vos, A.M., and Kossiakoff, A.A., The X-ray structure of a growth hormone–prolactin receptor complex, *Nature* **372,** 478–481 (1994); *and* Kossiakoff, A.A., Somers, W., Ultsch, M., Andow, K., Muller, Y.A., and De Vos, A.M., Comparison of the intermediate complexes of human growth hormone bound to the human growth hormone and prolactin receptors, *Protein Sci.* **3,** 1697–1705 (1994).

Sun, P.D. and Davies, D.R., The cystine-knot growth-factor superfamily, *Annu. Rev. Biophys. Biomol. Struct.* **24,** 269–291 (1995).

Walter, M.R., Windsor, W.T., Nagabhushan, T.L., Lundell, D.J., Lunn, C.A., Zauodny, P.J., and Narula, S.K., Crystal structure of a complex between interferon-γ and its soluble high-affinity receptor, *Nature* **376,** 230–235 (1995).

(c) Heterotrimeric G Proteins

Coleman, D.E. and Sprang, S.R., How G proteins work: a continuing story, *Trends Biochem. Sci.* **21,** 41–44 (1996); *and* Neer, E.J. and Smith, T.F., G protein heterodimers: New structures propel new questions, *Cell* **84,** 175–178 (1996).

Lambright, D.G., Sondek, J., Bohm, A., Skiba, N.P., Hamm, H.E., and Sigler, P.B., The 2.0 Å crystal structure of a heterotrimeric G protein, *Nature* **379,** 311–319 (1996).

Mixon, M.B., Lee, E., Coleman, D.E., Berghuis, A.M., Gilman, A.G.,and Sprang, S.R., Tertiary and quarternary structural changes in $G_{i\alpha 1}$ induced by GTP hydrolysis, *Science* **270,** 954–960 (1995).

Sondek, J., Bohm, A., Lambright, D.G., Hamm, H.E., and Sigler, P.B., Crystal structure of a G_A protein βγ dimer at 2.1 Å resolution, *Nature* **379,** 369–374 (1996).

Sondek, J., Lambright, D.G., Noel, J.P., Hamm, H.E., and Sigler, P.B., GTPase mechanism of Gproteins from the 1.7-Å crystal structure of transducin α-GDP·AlF$_4^-$, *Nature* **372,** 276–279 (1994).

Wedegaertner, P.B., Wilson, P.T., and Bourne, H.R., Lipid modifications of trimeric G proteins, *J. Biol. Chem.* **270,** 503–506 (1995).

Wall, M.A., Coleman, D.E., Lee, E., Iñguez-Lluhi, J.A., Posner, B.A., Gilman, A.G., and Sprang, S.R., The structure of the G protein heterotrimer $G_{i\alpha 1}\beta_1\gamma_2$, *Cell* **83,** 1047–1058 (1995).

(d) Bacterial Toxins

Bell, C.E. and Eisenberg, D. Crystal structure of diphtheria toxin bound to nicotinamide adenine dinucleotide, *Biochemistry* **35,** 1137–1149 (1996); *and* Weiss, M.S., Blanke, S.R., Collier, R.J., and Eisenberg, D., Structure of the isolated catalytic domain of diphtheria toxin, *Biochemistry* **34,** 773–781 (1995).

Merritt, E.A., and Hol, W.G.J., AB$_5$ toxins, *Curr. Opin. Struct. Biol.* **5,** 165–171 (1995).

(e) Receptor Tyrosine Kinases and Tyrosine Kinase–Associated Receptors

Hubbard, S.R., Wei, L., Ellis, L., and Hendrickson, W.A., Crystal structure of the tyrosine kinase domain of the human insulin receptor, *Nature* **372,** 754 (1994); *and* McDonald, N.Q., Murray-Rust, J., and Blundell, T.L., The first structure of a receptor tyrosine kinase domain: a further step in understanding the molecular basis of insulin action, *Structure* **3,** 1–6 (1995).

Songyang, Z. and Cantley, L.C., Recognition and specificity in protein tyrosine kinase-mediated signalling, *Trends Biochem. Sci.* **20,** 470–475 (1995).

(f) Ras-Mediated Signaling

Buergering, B.M.T. and Bos, J.L., Regulation of Ras-mediated signalling: more than one way to skin a cat, *Trends Biochem. Sci.* **20,** 18–22 (1995).

Cobb, M.H. and Goldsmith, E.J. How MAP kinases are regulated, *J. Biol. Chem.* **270,** 14843–14846 (1995).

Goldberg, J., Huang, H., Kwon, Y., Greengard, P., Nairn, A.C., and Kuriyan, J., Three-dimensional structure of the catalytic subunit of protein serine/threonine phosphatase-1, *Nature* **376,** 745–753 (1995).

Ihle, J.N, STATs: Signal tranducers and activators of transcription, *Cell* **84,** 331–334 (1996); Ihle, J.N., Cytokine receptor signaling, *Nature* **377,** 591–594 (1995); *and* Ihle, J.N. and Kerr, I.M., Jaks and Stats in signaling by the cytokine receptor superfamily, *Trends Genet.* **11,** 69–774 (1995).

Inglese, J., Koch, W.J., Touhara, K., and Lefkowitz, R.J., $G_{\beta\gamma}$ interactions with PH domains and Ras–MAPK signaling pathways, *Trends Biochem. Sci.* **20,** 151–156 (1995).

Karin, The regulation of AP-1 activity by mitogen-activated protein kinases, *J. Biol. Chem.* **270,** 16483–16486 (1995).

Maignan, S., Guilloteau, J.-P., Fromage, N., Arnoux, B., Becquart, J., and Ducruix, A., Crystal structure of the mammalian Grb2 adaptor, *Science* **268,** 291–293 (1995); *and* Guruprasad, L., Dhanaraj, V., Timm, S., Blundell, T.L., Gout, I., and Waterfield, M.D., The crystal structure of the N-terminal SH3 domain of Grb2, *J. Mol. Biol.* **248,** 856–866 (1995).

Nassar, N., Horn, G., Herrmann, C., Scherer, A., McCormick, F., and Wittinghofer, A., The 2.2 Å crystal structure of the Ras-binding domain of the serine/threonine kinase c-Raf1 in complex with Rap1A and a GTP analogue, *Nature* **375,** 554–560 (1995).

Pawson, T., Protein modules and signaling networks, *Nature* **373,** 573–580 (1995). [A review.]

Schindler, C. and Darnell, J.E., Jr., Transcriptional responses to polypeptide ligands: The JAK-STAT pathway, *Annu. Rev. Biochem.* **64,** 621–651 (1995).

Teresawa, H., Structure of the N-terminal SH3 domain of GRB2 complexed with a peptide from the guanine nucleotide releasing factor Sos; *and* Goudreau, N., Cornille, F., Duchesne, M., Parker, F., Tocqué, B., Garbay, C., and Roques, B.P., NMR structure of the N-terminal SH3 domain of GRB2 in its complex with a proline-rich peptide from Sos, *Nature Struct. Biol.* **1,** 891–897 *and* 898–907 (1994).

van der Greer, P. and Pawson, T., The PTB domain: a new protein module implicated in signal transduction, *Trends Biochem. Sci.* **20,** 277–285 (1995). [The phosphotyrosine-binding domain.]

Wu, X., Knudsen, B., Feller, S.M., Zheng, J., Sali, A., Cowburn, D., Hanafusa, H., and Kuriyan, J., Structural basis for the specific interaction of lysine-containing proline-rich peptides with the N-terminal SH3 domain of c-Crk, *Structure* **3,** 215–226 (1995).

Zhang, J., Zhang, F., Ebert, D., Cobb, M.H., and Goldsmith, E.J., Activity of the MAP kinase ERK2 is controlled by a flexible surface loop, *Structure* **3,** 299–307 (1995).

(g) Protein Tyrosine Phosphatases

Barford, Protein phosphatases, *Curr. Opin. Struct. Biol.* **5,** 728–734 (1995); *and* Barford, D., Jia, Z., and Tonks, N.K., Protein tyrosine phosphatases take off, *Nature Struct. Biol.* **2,** 1043–1053 (1995).

Jia, Z., Barford, D., Flint, A.J., and Tonks, N.K., Structural basis for phosphotyrosine peptide recognition by protein tyrosine phosphatase 1B, *Science* **268,** 1754–1758 (1995).

(h) Plekstrin Homology Domain

Ferguson, K.M., Lemmon, M.A., Schlessinger, J., and Sigler, P.B., Crystal structure at 2.2 Å resolution of the pleckstrin homology domain from human dynamin, *Cell* **79,** 199–209 (1994); *and* Timm, D., Salim, K., Gout, I., Guruprasad, L., Watersfield, M., and Blundell, T., Crystal structure of the pleckstrin homology domain from dynamin, *Nature Struct. Biol.* **1,** 782–788 (1994).

Ferguson, K.M., Lemmon, M.A., Schlessinger, J., and Sigler, P.B., Structure of the high affinity complex of inositol triphosphate with a phospholipase C pleckstrin homology domain, *Cell* **83,** 1037–1046 (1995).

Sareste, M. and Hyvönen, M., Pleckstrin homology domains: a fact file, *Curr. Opin. Struct. Biol.* **5,** 403–408 (1995).

Shaw, G., The pleckstrin homology domain: an intriguing multifunctional protein module, *BioEssays* **18,** 35–46 (1996).

(i) Phosphoinositide Cascade

Newton, A.C., Protein kinase C: structure, function, and regulation, *J. Biol. Chem.* **270,** 28495–28498 (1995).

Zhang, G., Kazanietz, M.G., Blumberg, P.M., and Hurley, J.H., Crystal structure of the Cys2 activator-binding domain of protein kinase Cδ in complex with phorbol ester, *Cell* **81,** 917–924 (1995).

(j) Neurotransmission

Harel, M., Kleywegt, G.J., Ravelli, R.B.G., Silman, I., and Sussman, J.L., Crystal structure of an acetylcholinesterase–fasciculin complex: interaction of three-fingered toxin from snake venom with its target, *Structure* **3**, 1355–1366 91995); *and* Bourne, Y., Taylor, P., and Marchot, P., Acetylcholinesterase inhibition by fasciculin: crystal structure of the complex, *Cell* **83**, 503–512 (1995).

Jaffrey, S.R. and Snyder, S.H., Nitric oxide: a Neural messenger, *Annu. Rev. Cell Dev. Biol.* **11**, 417–440 (1995).

Kwong, P.D., McDonald, N.Q., Sigler, P.B., and Hendrickson, W.A., Structure of β_2-bungarotoxin: potassium channel binding by Kunitz modules and targeted phospholipase action, *Structure* **3**, 1109–1119 (1995).

Unwin, N., Acetylcholine receptor channel imaged in the open state, *Nature* **373**, 37–43 (1995).

NOTES

NOTES

NOTES

NOTES

NOTES

NOTES

NOTES

NOTES

NOTES

NOTES

NOTES